A political geography of
AFRICA

A political geography of
AFRICA

E. A. BOATENG

formerly Professor of Geography at the
University of Ghana

CAMBRIDGE UNIVERSITY PRESS

CAMBRIDGE

LONDON · NEW YORK · MELBOURNE

Published by the Syndics of the Cambridge University Press
The Pitt Building, Trumpington Street, Cambridge CB2 1RP
Bentley House, 200 Euston Road, London NW1 2DB
32 East 57th Street, New York, NY 10022, USA
296 Beaconsfield Parade, Middle Park, Melbourne 3206, Australia

© Cambridge University Press 1978

First published 1978

PRINTED IN GREAT BRITAIN BY
COX & WYMAN LTD, LONDON, FAKENHAM AND READING

Library of Congress Cataloguing in Publication Data
Boateng, E A
A political geography of Africa
Bibliography: p.
Includes index.
1.   Africa—Politics and government.
2.   Geography, Political.   I. Title.
DT31.B57   320.9′6   77-80828
ISBN 0 521 21764 4 hard covers
ISBN 0 521 29269 7 paperback

To my wife Evelyn

Contents

List of maps

Foreword

As a branch of geography, political geography has never flourished in Britain or the Commonwealth as much as might have been expected. This is particularly surprising in view of the fact that one of the subject's great pioneers in Britain – the first Reader in Geography in the University of Oxford, Sir Halford J. Mackinder – was the author of a very significant book published towards the end of the First World War, *Democratic Ideals and Reality*, and was for several years a Member of Parliament. During the 1920s, British geography was considerably influenced by the comprehensive book, *The New World: Problems in Political Geography* (1921), written by an American, Isaiah Bowman, who had been closely associated with the American delegation to the Versailles Peace Conference; while a few years later C. B. Fawcett's *A Political Geography of the British Empire* (1933) might have laid the foundation for studies in depth of that remarkable institution, the British Empire, whose influence and role in world affairs were for a time at least so very important.

Then the Second World War broke out and was followed by a period during which the Empire was increasingly replaced by a Commonwealth, in which all but a minority of states enjoyed political – though not necessarily economic – independence. Meanwhile the range of geography was widening and the approach was being modified. But of the much greater number of professional geographers, only a relatively small, though distinguished, group has concentrated its attention upon political geography, despite the undisputed importance of what the author of this book calls 'the study of those aspects of the geography of politically organised areas that are relevant to the existence and effective functioning of such areas both internally and in their external relationships'. Very few focused their attention upon the continent of Africa although two of them, B. W. Hodder and D. R. Harris, edited *Africa in Transition: Geographical Essays* (1967); but useful and stimulating though the contributions to the volume undoubtedly are, the two editors and their three collaborators were 'expatriates' and therefore tended to express the outsiders', as opposed to an African, point of view.

With the emergence of Africa from colonialism to independence during this second half of the twentieth century there has been a real need for an over-view of the whole continent as it is in the 1970s; and it is important that such an assessment should be undertaken by one person and that he or she should be an African. Few could be better qualified for this task than Professor E. A. Boateng, a Ghanaian geographer with a very considerable academic record and a person with extensive administrative, and indeed political, experience. My acquaintance with Amano Boateng goes back to 1946 when I first met him in Ghana (then the Gold Coast) shortly before he came to Britain to read geography in the University of Oxford. I was his tutor and he selected as his Special Subject the one for which I was responsible, 'the social and political geography of British tropical Africa and the Caribbean'. Subsequently I was his supervisor for the research degree of B. Litt. On returning home he joined the staff of the newly created Department of Geography in what was then the University College of the Gold Coast and rose to become Professor and Head of the Department of Geography in the University of Ghana before moving to the University of Cape Coast as Vice-Chancellor. Since 1973 he has been Executive Chairman of the Environmental Protection Council of Ghana.

Despite his many administrative and other preoccupations Professor Boateng has managed to maintain close contact with geography and has enriched his understanding of its practical aspects through extensive foreign travel in Africa and elsewhere as his country's representative on a number of official delegations to important international conferences and meetings, including the United Nations General Assembly. Thus he is in a particularly favourable position to see his own continent in its global context and to appreciate the political problems of its many diverse countries in their continental setting.

Not everyone will agree with everything that the author has written in this book. But what is important is that he expresses a point of view and sustains an argument about African affairs in general and in depth that must be respected because of his academic approach and his very wide experience of affairs both in his own country of Ghana and in Africa as a whole. He brings to these studies the trained mind of a geographer, and he paints on a broad canvas with due attention to detail as and when necessary. I am honoured to be asked by the author to write a foreword for his book and am happy to commend it to a wide range of readers who will be helped by it to a greater understanding of the complexities of the African scene and to a deeper knowledge of the challenges which this most fascinating of the continents presents to the world at large.

September 1977   Robert W. Steel
                                   Principal, University College of Swansea;
                                   formerly Fellow of Jesus College, Oxford,
                                   and John Rankin Professor of Geography,
                                   University of Liverpool

Preface

One of the most fascinating yet bewildering features of the contemporary African political scene is the rapid rate of change all over the continent and its generally unpredictable character. This presents anyone attempting to write a book on the political geography of the continent with the very difficult problem of choosing a suitable time-frame from his work and deciding how best the essential features of the ever-changing kaleidoscope of events can be captured and conveyed to the reader in a coherent and intelligible form, since nothing that is written about the present seems likely to remain valid for any great length of time. Recognising the problems posed by this state of affairs, what I have tried to do in this book has been to select what I consider to be the really salient political and socio-economic problems and issues facing the continent and its various regions and countries today and to examine them within their geographical and historical context in the hope that this will facilitate an understanding of their real import and provide a reasonably sound basis for assessing future developments and trends.

Owing to the immense size of Africa and the great diversity and complexity of its problems, I have felt obliged to be highly selective in the choice of topics and issues for discussion, especially in Part III of the book, which deals with the individual states and regions of the continent, the nature of the topics and issues selected depending on the special circumstances of each particular area. This is in conformity with the definition of political geography which I give in the book as the study of those aspects of the geography of politically organised areas that are relevant to the existence and effective functioning of such areas both internally and in their external relationships. I feel that to have attempted a wider coverage of material would not only have made my task immeasurably more difficult but would have presented the reader with an account far too lengthy and detailed to provide the focus and perspective necessary for a proper understanding of the really significant problems. I have no doubt that with the passage of time new problems and new issues will emerge, but it is my hope that this will serve to underscore rather than diminish the relevance of the facts and viewpoints set forth in this book.

The book is divided into three parts. Part I deals with the general principles of political geography, Part II with certain broad topics which constitute an essential background to an understanding of the politico-geographical problems of the African continent, while Part III examines in some detail what are considered to be the major problems and issues facing the individual states and regions of the continent in the field of political geography.

The idea of writing this book first suggested itself to me some twenty years ago when I began teaching a course on the political geography of Africa as one of the options for the honours degree in geography at what is now the University of Ghana. The idea took more concrete shape in the 1965–66 academic year when I had the good fortune to be awarded a Smuts Visiting Fellowship in the University of Cambridge with the freedom to research and write in any area of my choice. The tenure of a visiting professorship in the University of Pittsburgh, USA, during the summer trimester of 1966 gave me further opportunities for research and, even more important, for trying out some of my findings and ideas on various groups of highly enthusiastic but refreshingly critical students taking my course on the political geography of Africa. The response I obtained from these students and, later, from my own students at the University of Ghana further strengthened my conviction that there was a need for the kind of book I was planning to write, and but for a number of quite unexpected administrative responsibilities that subsequently came my way I might have completed the work long before now. But in a way the delay has proved beneficial; for it has allowed time for many of the important political developments in the continent which were then only in their initial stages to unravel themselves, thus making it easier to discern their implications and trends.

In writing this book I have derived invaluable assistance from a large number of people and organisations to all of whom I feel deeply grateful, although only a few of them can be mentioned here by name owing to limitations of space. First and foremost, I wish to thank the Trustees of the Smuts Memorial Fund in Cambridge for making it possible for me to embark on the initial research for the work, free from the preoccupations of teaching and administration. Next, I must express my appreciation to the staffs of the various libraries, especially the Cambridge University Library, the University of Pittsburgh Library, the Balme Library, University of Ghana, and the Padmore Research Library in Accra, where I did the bulk of my research, for the assistance they gave me with the collection of my material.

Of the many academic colleagues who gave me the benefit of their knowledge and advice or assisted me in other ways I should like to make special acknowledgement to Professor R. J. Harrison Church of the

London School of Economics, Professor George Benneh of the University of Ghana and Professor Hibberd V. B. Kline, Jr, Chairman of the Department of Geography in the University of Pittsburgh. Above all, however, I feel I owe a special debt of gratitude to Professor R. W. Steel, Principal of the University College of Swansea, and formerly my tutor at Oxford, who gave me my first insight into the political geography of Africa at a time when the idea of political independence for Africa, especially black Africa, was as yet only a distant dream. He made many valuable comments on my manuscript and has placed me further in his debt by agreeing to write the Foreword to this book.

All the maps contained in the book have been drawn by Mr J. F. Antwi of the National Atlas Project of Ghana with the assistance of other members of the cartographic staff. I feel all the more grateful to them because of the tremendous pressure under which they had to work in order to complete the task on time. I feel especially indebted to my Private Secretary in the Environmental Protection Council, Miss Margaret Warden, as well as to Mr Joseph Adjei also of the same Council and to my former Secretary in the Vice-Chancellor's Office at the University of Cape Coast, Mr John Kuofie, for typing my manuscript during its various stages.

Finally, I wish to express my profound gratitude to my wife and our daughters for their forbearance and understanding throughout the long period during which this book has been in preparation. But for their cooperation and encouragement I know that I could never have accomplished the task.

Accra
February 1977

E. A. Boateng

PART I

Principles of political geography

I

The meaning and scope of
political geography

Political geography as understood today is concerned with the study of
those geographical aspects of politically organised areas that affect the
political life and activities of such areas. In most writings on political
geography the unit that is most commonly employed as a basis for study
is the state; but in fact any unit, whether large or small, that is
politically organised can serve as a legitimate basis for study.

The German geographer, Friedrich Ratzel, who lived between 1844
and 1904, is generally regarded as the founder of political geography in
the modern sense, and his book, *Politische Geographie*, published in 1897
is considered as the first modern treatise on the subject. Ratzel did not
offer a precise definition of the subject, but he expressed the view that
what was known as political geography in his day was nothing more
than what we would today describe as 'regional geography'. He held
that this kind of geography could be raised to a higher plane if more
attention were given to the relations between the state and the land or
space which it occupied. For Ratzel this relationship between the state
and its land area (German: *Raum*) was the real essence of political
geography. He saw the state and the land area it occupied as bound
together in a close organic relationship in which the land offered the
state certain opportunities for growth and expansion, while it was the
duty of the state to grasp the meaning and range of these opportunities
and utilise them to the full.

The relationships between states and the land they occupy are, of
course, infinite in their range. But Ratzel did not concern himself with
all of them: the ones that particularly engaged his attention were those
that tended to the enhancement of the state and the fulfilment of its
political destiny. It is obvious from this that what Ratzel understood by
political geography was not the subject as we understand it today but
rather the study of the territorial basis of national power.

Apart from its rather limited scope, Ratzel's interpretation of political geography led to some unfortunate developments, especially in his
own country, Germany. The idea that states have a destiny inherent in
their geography which it is their duty to recognise and fulfil led to an
excessive preoccupation among certain of his countrymen, especially

military leaders and politicians, with the destiny of Germany and the opportunities which its geography offered for its fulfilment. And since their principal objective at the time was national power, they tended to place excessive emphasis on those aspects of political geography which supported their desire for territorial expansion or, as Adolf Hitler subsequently called it, *Lebensraum*, that is, living space.

Despite the shortcomings in Ratzel's conception of political geography and the voluminous literature that has appeared on the subject since his day, the importance of his pioneering efforts still remains unchallenged.

Within the present century many writers have produced works on political geography or on special facets of the subject. It would be impossible to mention all of them, but such names as Halford Mackinder, Isaiah Bowman and Derwent Whittlesey are bound to stand out in any list. Of these by far the most outstanding, without any doubt, is the American geographer, Isaiah Bowman, who in the opinion of Harriet Wanklyn (Wanklyn 1961) is the only twentieth-century geographer to have had anything approaching Ratzel's grasp of the field. Like Ratzel, Bowman did not in any of his writings offer an explicit definition of political geography, but in his classic work, *The New World*, first published in 1921, he gave a clear enough indication of what he understood by the term by his general method of approach and his selection of topics. His really important contribution in this respect was that, unlike Ratzel and his disciples, he concerned himself with an objective analysis of contemporary world problems as seen in the light of geography, thus underlining the very important point that what constitutes the political geography of any area depends upon the particular problems and circumstances of the area concerned within the context of the prevailing local, national or international situation.

Many subsequent writers since Bowman's day have sought to define the subject in rather more explicit terms. Derwent Whittlesey, for example, another of the outstanding American pioneers, stated that the essence of political geography is 'the differentiation of political phenomena from place to place over the earth' (Whittlesey 1944), while Richard Hartshorne, who is probably the best known living exponent of the subject in America, defined it as 'the study of areal differences and similarities in political character as an interrelated part of the total complex of areal differences and similarities', and went on to state that 'the interpretation of areal differences in political features involves the study of their interrelations with all other relevant areal variations, whether physical, biotic or cultural in origin' (Hartshorne 1954). Six years later he restated this definition more simply and lucidly and described political geography as 'the study of the variations of political phenomena from place to place in interconnection with variations in other features of the earth as the home of man' (Hartshorne 1960).

British geographers have also offered a variety of definitions. In his book, *The New Europe*, written in 1945, Walter Fitzgerald put forward the view that the 'objective of political geography is an investigation of the extent to which the nature of States, together with their organization and interrelations, is influenced by, and adjusted to, conditions of geography'. A. E. Moodie in his book, *Geography Behind Politics*, stressed the importance of the internal and external relationships of states and the need to harmonise them if the state is to function successfully, while S. W. Wooldridge and W. G. East in their important book, *The Spirit and Purpose of Geography*, published in 1951 expressed the view that 'political geography focuses attention on both the external and internal relations of states'. But perhaps the most explicit definition of all is that put forward by Norman J. G. Pounds, who states in his book, *Political Geography*, published in 1963, that 'political geography is concerned with politically organised areas, their resources and extent, and the reasons for the particular geographical forms which they assume'.

Varied as these definitions are, most of them appear to agree with Ratzel's view that political geography involves a study of the relationships between the state as a political phenomenon, on the one hand, and geographical conditions and circumstances, on the other. However, not all of them refer to the State as the focal point of study; rather they talk in terms of areas or places. Although there is no doubt that the State forms the supreme political entity in all parts of the world, there exist other entities of a political nature which also play a significant role in human affairs. All such entities, whether they be tribes, cantons, countries, electoral districts, constituencies, provinces or regions, constitute valid political units capable of study within an identifiable geographical milieu, and any definition of political geography that aims at completeness must take them into account. This is why the term 'politically organised areas' is preferable to the word 'States' as a description of the basic unit upon which the study of political geography is based.

Political geography is not concerned solely with the relationships between politically organised areas and geographical conditions. Every politically organised area, however small, forms part of a larger whole, whether or not its inhabitants recognise the fact.

No politically organised area, however small or however parochial it may wish to be, can live in complete isolation; hence the need to extend our field of enquiry in political geography to the relationships between different political entities from the smallest to the largest and the way in which these relationships are influenced by geographical conditions. These external relationships are especially important in the present

5

world because of the tremendous advances in transport and communications which have brought all parts of the world much closer together than at any previous time in recorded history. The study of international relations has now become a very important aspect of political geography, and consequently organisations like the United Nations, which are concerned to contain, reduce or eliminate conflict and promote harmonious relations between different states and nations, must necessarily engage the attention of political geographers.

Having established the nature of the political entities or units which form the subject matter of political geography, we need next to consider the nature of the relationships between these entities and the geographical conditions to which the political geographer must give his attention. As we observed earlier when discussing Ratzel's definition of the subject, these relationships are infinite in their range, and it is necessary to decide just which of them are truly relevant.

No two politically organised areas are exactly alike in their political conditions and circumstances. Similarly, in no two cases shall we find the same geographical factors operating in equally decisive ways in the political life of such areas. In other words, the criteria for selecting the geographical aspects to be considered in any study of political geography must vary from one area to the next, depending on the peculiar circumstances of each particular area at any given time. This is the outstanding lesson to be learnt from Isaiah Bowman's *The New World*, where the treatment of each area is unique. For what the political geographer must aim to do is not to make a stereotyped study or analysis in which all the geographical features alike simply provide a uniform backdrop without any attempt to distinguish between those which are really significant and those which are not, but rather to examine those aspects of the geography of the area in question which significantly affect its political life, having regard to the need for every politically organised area to achieve viability and to maintain its territorial integrity.

Taking all the foregoing points into consideration, we may define political geography as the study of those aspects of the geography of politically organised areas which are relevant to the existence and effective functioning of the areas concerned as political entities both internally and in their external relationships.

Apart from its simplicity, this definition has the merit of being sufficiently broad and general to accommodate any scale of politico-geographical investigation. It is also valid whether we are considering purely local and internal problems or whether we are concerned with the wider problems of international politics involving several countries or indeed the whole world.

Although any politically organised area qualifies as a subject for study in political geography, in practice the units that are most commonly considered are states and countries. This is not surprising. After all, these are the units with which the exercise of political power, especially sovereign political power, in its most complete and obvious forms is associated.

In studying political geography, our principal concern must be with the territorial rather than with the purely political aspects. Every state, as Ratzel pointed out, consists of people and a definite space or territory which they inhabit. The way in which the people utilise their territorial base depends very largely on their social and political organisation, while the nature of the territorial base can influence in important ways the mode of life of the people and their outlook.

It is obvious that the student of political geography must have some acquaintance with political science, for he must be familiar with the way in which political entities function. But first and foremost he must be a geographer, capable of relating political activities and needs to geographical circumstances, which collectively form the milieu in which political entities actually exist and function. Within geography itself, political geography belongs more to the human than to the physical branch, and it leans heavily on most of the subdivisions of the human branch, in particular economic and historical geography. Economic geography is important because political life is so very closely connected with economic activity in the modern world, while historical geography is relevant because most political and associated phenomena cannot be adequately understood except in relation to the past.

In most modern works on political geography the word *'geopolitics'* invariably crops up, and indeed certain writers, especially in America, tend to use the term as if it was synonymous with political geography. This is a mistake; for the two terms mean quite different things, although they bear a certain relationship to one another. The originator of the term 'geopolitics' was Rudolf Kjellen (1864–1922). Kjellen was a professor of political science at Uppsala and a close follower of Ratzel. The actual word he used was 'Geo-politik', which he defined as the science of the state as a realm in space (Kish 1942). Ratzel had envisaged the state as a natural organism capable of growth and expansion in accordance with simple organic laws, but Kjellen went a step further and suggested that in seeking territorial expansion the state was at liberty to employ modern cultural advances and techniques in the achievement of its desired goals. He argued that the final objective of a state in its development of power was 'to acquire good natural frontiers externally, and harmonious unity internally'.

Kjellen saw geopolitics as one of the five basic divisions of political science, the others being: (1) Krato-politik, the science of the legal

7

organisation of the power of the state, (2) Demo-politik, the forms of the political organisation of the masses, (3) Oeco-politik, the science of the forms of the production and consumption of goods, and, (4) Socio-politik, the science of the social organisation of the state.

Geopolitics was obviously a ready-made tool in the hands of any state intent on territorial expansion and national power, and it is not surprising that it was enthusiastically espoused by Germany between the First and Second World Wars. The man responsible for popularising geopolitics in Germany was Karl Haushofer (1869–1946).

Haushofer was a gifted scholar, who early in life abandoned an academic career for a military one. Later, after travels in Japan and the far East, he taught geography at Munich and subsequently rose during the First World War to the rank of major-general in the German army. After the armistice he returned to Munich to teach political geography and military science and in 1924 helped to found an influential journal known as *Zeitschrift fur Geopolitik*, which was devoted to the furthering of geopolitics.

It is said that Haushofer was influenced very greatly in the development of his ideas by the British geographer and statesman, Sir Halford J. Mackinder (1861–1947), who was almost an exact contemporary of his. Mackinder was responsible for two very important publications on political geography. The first was a paper entitled 'The Geographical Pivot of History', which was read to the Royal Geographical Society in 1904 and made a great impact on the geographical world at the time; the second was a book entitled *Democratic Ideals and Reality* published in 1919 in which he underlined and developed further the ideas contained in the earlier article. Mackinder was one of the first geographers to appreciate the role of space relationships in the development and interplay of national power, and such was the sweep and force of his arguments that it is not surprising that his work should have had so much influence on Haushofer. However, his sole concern was to make an analysis and a prognostication based on history and geography and not to advance the cause of any one nation or group of nations.

It is noteworthy that Haushofer himself considered political geography to be quite distinct from geopolitics. In his view, 'political geography represents the science of the distribution of political power over the different regions of the world and the conditions of political power by, and its dependence on, surface features, climate and other aspects of physical geography'. In contrast to this, he saw geopolitics as essentially dynamic, 'a way of educating the masses in the concept of space' (Kish 1942).

It would seem, therefore, that what Haushofer defined as 'geopolitics' was almost identical with what Ratzel had earlier described as political geography, except that Haushofer bent the subject much more obviously and directly to the needs of his own fatherland, Germany.

Geopolitics was greatly exploited by Hitler before and during the Second World War in the pursuit of his territorial ambitions, and consequently the word and everything it stood for became greatly repugnant to scholars in the rest of Europe, especially in Britain and France. Thus, although it is quite possible that under normal circumstances the word *geopolitics* might gradually have replaced the term *political geography* as being more impressive and less unwieldy, especially in its adjectival form *'geopolitical'* as contrasted with *'politico-geographical'*, a sharp difference has come to exist between them.

Political geography is considered in respectable geographical circles as an objective study based on the examination and analysis of facts as they truly are, while geopolitics is regarded as a tendentious and biased study, a distortion and prostitution of political geography which is quite unworthy of serious scholars. Professor E. G. R. Taylor described it as 'political geography charged with emotion' (Taylor 1945), while the American geographer, H. Weigert, described it as 'the application of geographical principles ... in the game of power lusts' (Weigert 1942). The French geographer, A. Demangeon, talking specifically about German geopolitics, observed, perhaps with excessive though understandable acerbity, that it had deliberately rejected the scientific spirit in every way and become an instrument of war (Demangeon 1932). In view of all this, it is essential not to confuse geopolitics with political geography.

Nevertheless, there is still a need in the vocabulary of political geography for a word which conveniently describes those aspects of world strategy and of space relationships whose special significance lies in the fact that the earth is a globe and not a flat surface. Mackinder's writings clearly brought out this global concept of terrestrial space and its importance in world military strategy, athough he did not offer a specific word for conveying the meaning of the concept. Immediately after the Second World War, when the widespread use of air transport began to create a new awareness of the global character of the earth and its strategic implications, a number of writers, especially in America, attempted to revive the use of the word *geopolitics* for the purpose of conveying this global concept, but without much success, no doubt on account of the earlier unfortunate historical connotations of the word. Perhaps a more acceptable, and certainly more explicit, word might be the word *geo-strategy* as employed by Blouet in a recent article on Mackinder (Blouet 1976).

9

2

Elements of political geography

As has been shown in the preceding chapter, political geography is concerned with the study of politically organised areas. Of all the many different kinds of politically organised areas to be found in the world the state is the most important and the most widely recognised in international terms. Every state consists of five basic elements: (1) a territorial base, (2) a resident population, (3) a system of government, (4) an economic base, and (5) a system of transport and communications. Each of these elements will now be considered in turn.

TERRITORIAL BASE

Territory is the most fundamental of the five elements of the state. Without territory no state can claim existence. In simple terms the word territory means an expanse of land, whether large or small.

Location

Every territory or territorial base must have a definite location as well as boundaries which define its physical limits. Location may be given in terms of latitude and longitude. It may also be given in terms of latitudinal zones, e.g. the equatorial zone lying between 5 and 10 degrees of the equator, the two tropical zones lying between these limits and the tropics of Cancer and Capricorn, i.e. $23\frac{1}{2}°$N and $23\frac{1}{2}°$S, the two temperate zones or middle latitudes, as they are sometimes called, between the tropics and the Arctic and Antarctic Circles, i.e. $66\frac{1}{2}°$N and $66\frac{1}{2}°$S, and finally the two polar zones lying between these Circles and the poles. This method of showing location is of particular geographical importance because of its strong climatic undertones and the basis it provides for making various inferences regarding geographical conditions in any particular place.

The location of places by reference to longitude is of little use here, since few geographical phenomena except for the incidence of day and night depend on it. More significant is the division of the world into an eastern and a western hemisphere based on the longitudinal arrangement of the continents. But an even more practical division of the world today is into the western world, the eastern bloc and the third world or the developing countries.

While location in global terms is undoubtedly important, the political fortunes of states are often affected much more directly by their own size, shape and physical characteristics and their spatial relationships with their immediate neighbours. All states need to ensure the safety and integrity of their territory, and most require access to their neighbours for purposes of trade and communication. For this reason every aspect of location that conduces to easy defence as well as to reasonable access to neighbouring states is important.

Accessibility is very closely related to the available means of transport. Practically all states are able to employ one or other of the three principal means of transport available to man: land, water and air transport. While states with access to the sea can have uninhibited and often easy access to many other states owing to the time-honoured and established 'freedom of the sea', states which are landlocked cannot communicate either by land, or by air or by water with any other states, except their immediate neighbours, without the consent and goodwill of all the intervening states, since there is as yet no 'freedom of the air' and every state can lay claim to the air space immediately above its territory. That is why the question whether a state has an insular, continental or maritime location and whether it lies on or comprises a peninsula, an isthmus or an archipelago, etc. is of such importance in political geography, and terms like *enclave, exclave, glacis, corridor*[1] all form part of the vocabulary of political geography (Fig. 1).

Another aspect of location is the extent to which it confers special advantages for obtaining certain important natural resources either locally or from external sources. It is well-known that the distribution of certain natural resources, such as agricultural, vegetable and animal resources, is broadly related to latitudinal position and sometimes to altitude. But even here all kinds of irregularities supervene on account of irregularities in the sizes and spatial distribution of the continents and of the various factors responsible for climate and vegetation. When it comes to the distribution of minerals, however, location in terms of latitude or longitude is almost totally irrelevant, except in the case of a very small number of minerals, such as bauxite, whose

---

1 enclave – A state or an outlying portion of a state wholly surrounded by the territory of another state. The term is used from the point of view of the surrounding state.

exclave – The same as 'enclave' but as seen from the point of view of the parent state.

corridor – A narrow strip of territory belonging to one state which interrupts the territory of another state in order to give access to a port or some other geographical feature or facility of importance, e.g. a river or coastline.

glacis – A relatively small portion of a state which extends across a mountain divide forming the boundary between that state and another.

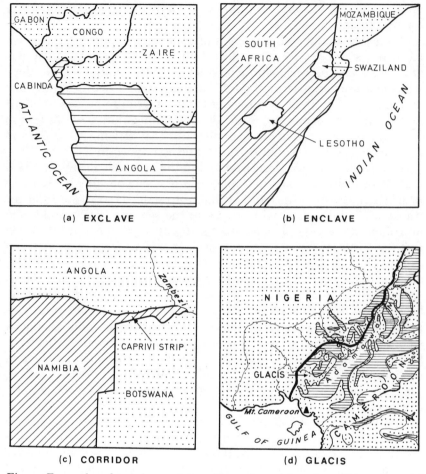

Fig. 1. Examples of certain special configurations: (a) exclave, (b) enclave, (c) corridor, (d) glacis

occurrence can be linked directly or indirectly to certain climatic conditions.

Size and shape

Size and shape are two other important aspects of territory. Other things being equal, the larger a state is the larger is its resource base. Even where the population is high and the population–resource ratio is no different from that of a small state with a lower population, the large state still has an advantage in that its *geographical momentum* is greater, that is, the sum total of its geographical resources and their resulting impact is greater.

There are over 120 states in the world today, ranging in size from the Vatican City State in Rome with an area of about 0.43 sq. km (one-sixth of a square mile) to the Soviet Union with an area of about 22,292,000 sq. km (8,600,000 sq. miles). Among the independent states of Africa, too, wide contrasts occur, from The Gambia and Lesotho with areas of 11,295 sq. km and 30,355 sq. km (4,361 and 11,720 sq. miles) respectively to Sudan and the Republic of Zaire with areas of 2,505,813 and 2,345,409 sq. km (967,494 and 905,562 sq. miles) respectively.

Apart from economic considerations, size can also be an important factor in guaranteeing the security of a state. It is much easier to gain control of a small state in a short space of time than it is of a large; a large state also requires a much more complex machinery for its own internal control and organisation than a small one. However, given proper internal organisation, it is more difficult for all the vital parts of a large state to be seized or controlled from outside. Large states also have much longer lines of possible retreat within their own territory than smaller ones. This is an advantage which Russia demonstrated when Napoleon attacked it in 1812 and again when Hitler attacked it during the Second World War.

Both shape and size are important, and the perimeter of a state, that is, the length of its boundaries, is dependent upon both shape and size. The aspect of shape that is especially relevant in political geography is the extent to which it makes for compactness in the territorial base. The ideal shape for a state, other things being equal, is a circular one. This combines maximum area with minimum length of boundaries and confers an advantage for defence as well as for offence. It also ensures that the centre is equidistant from all the peripheral areas. France is often quoted as approximating this desirable compactness of form. At the opposite extreme is the state with an elongated or linear form approaching a narrow rectangle, such as The Gambia. Here, minimum area is combined with maximum length of boundaries and the advantages that go with a compact form are all reversed. Such a state is highly vulnerable, for it has a long front along which it is exposed to its neighbours, especially if it is wholly or almost wholly landlocked.

The degree of compactness possessed by any state can be expressed mathematically by the ratio between the square root of its area and its total perimeter or boundary, giving us the equation: $C = \sqrt{A}/B$, where C is the degree of compactness, A the area of the state and B the length of its boundary. In a state with the shape of a perfect circle the ratio of A to B, which may be termed the compactness ratio, is approximately 1:3.5. Since the circle is the figure with the maximum possible degree of compactness, it is evident that the more that the compactness ratio of any given state approaches the ratio 1:3.5, the more compact it can be said to be. But, of course, mere compactness is not everything, even

13

from a strategic or security point of view; a great deal depends on the nature of the boundary itself – in particular, whether it runs wholly on land or whether it borders the sea, either wholly or partly, and also whether the adjoining or neighbouring states are friendly or hostile.

In discussing the shape of the territorial base, it needs to be remembered that states are not just flat pieces of land. Until recently, states were thought of largely in terms of two dimensions, that is, in terms of their extent solely on the surface of the earth. In fact, however, the area which states control consists of much more than this; it is a three-dimensional concept. In theory, every state can claim the whole of the area lying beneath its soil down to the centre of the earth. With recent developments in the idea of 'air space' also, it can be assumed that states extend several miles above their land area into the atmosphere, though how far up is not definite. Perhaps here the definition needs to be pragmatic. The earth's atmosphere extends for about 500 km (300 miles), though the troposphere, which is the part that is really relevant to life on the earth, extends up to only about 8–18 km (5–11 miles) (Critchfield 1966). It can reasonably be supposed that every state is entitled to claim at least this portion of the atmosphere above its territory as its own. However, no state can insist on respect for its air space, whatever its extent, unless it is in a position to challenge effectively the passage of strangers within any part of this space. Rockets can circle the earth freely without evoking protests, but most conventional aeroplanes, which cannot go so high, need to respect the conventions of air space, which are being increasingly defined with greater precision.

Along the sea, an offshore distance of between 6 and 12 miles (10–19 km) is now generally recognised internationally as the limit of territorial waters within which maritime states can exercise exclusive control. But even here a growing number of states are now claiming much more than this, especially where valuable offshore fisheries or other important economic resources occur. Since the United Nations Law of the Sea Conference held in 1974, a new concept has emerged, namely, that of an economic zone of 200 miles (322 km) from the shore within which maritime states may enjoy exclusive rights of economic exploitation without the denial of rights of passage to other states except within the 12-mile zone (19.3 km) of territorial waters. No definite conclusion has as yet been reached on the idea and the issue has been further complicated by a proposal to accord certain fishing and other rights within the economic zone to neighbouring landlocked states. However, there are strong indications that this new concept will eventually gain the support of the international community, especially since a number of coastal states have already taken unilateral action to enforce their rights of exclusive economic exploitation within a zone of up to 200 miles of their offshore waters.

When the problem is viewed in this way, it is possible to regard the area of every state today as forming a kind of pyramidal or conical solid reaching from the surface to a point at the centre of the earth and up through the air to an area of much greater extent than the land surface at the outer limits of the atmosphere. This idea of the three-dimensional state is not a fanciful invention aimed at exciting the imagination; it is something which calls for serious consideration in view of the rapid scientific and technological advances being made by man in the exploration of outer space as well as of the earth's interior.

The value of territory as a base for human activity depends largely on the extent to which it lends itself to effective utilisation. Different kinds of terrain present different opportunities and different problems. Altitude and relief are perhaps the most important aspects of surface configuration. Human bodies are designed to function as near as possible to sea level, where atmospheric pressure and the consequent availability of oxygen are at their highest. Beyond a height of about 1,526 m (5,000 feet) above sea level man needs to develop certain special physiological adaptations. Relief is concerned not with altitude but rather with the differences between the highest and lowest points in any given area. If the pattern of relief is such that the surface is uneven, then the process of utilisation by man becomes more difficult.

Transportation and movement are always difficult in countries of bold relief, but at the same time it is well-known that the presence of mountains and rugged terrain can have important beneficial effects on climate and on the flow of surface water, which can be much more easily harnessed for the production of hydro-electricity.

If the surface configuration of a state is such that high mountains and rugged relief occur in the marginal areas, leaving a considerable area of fairly level land, then the state can derive all the benefits that mountainous terrain confers, including protection against troublesome neighbours, while at the same time having a reasonable area of low land for agricultural and other economic activities and for settlement.

Although the advantages of high and rugged terrain are often outweighed by their disadvantages, the history of many countries shows that such areas have again and again provided areas of refuge for people during times of trouble. Examples are the Central Massif of France, the Swiss Alps, the Chota Nagpur Plateau of India and the Jos Plateau of Nigeria. But perhaps the classic examples of all are Tibet and Ethiopia.

Boundaries

The boundaries of a state are among the most important of its attributes. If the internal part of a state is regarded as the vital skeletal frame, then its boundaries form the outer shell. It is along boundaries that the state impinges on its neighbours. If there is to be trade with one's neighbours, it is across boundaries that such trade must flow. Similarly, if there is invasion from outside, boundaries must necessarily be crossed. Herein lies the justification for Lord Curzon's oft-quoted observation that 'frontiers form a razor's edge on which hang suspended the modern issues of war or peace, of life or death for nations'.[2]

The terms *boundaries* and *frontiers* are often used as though they were synonymous and interchangeable. Strictly speaking, this is an error. Boundaries are the man-defined lines which mark the limits of states and determine the extent of their sovereign authority. On the other hand, frontiers refer to zones of separation and conform more to nature, which abhors sharp divisions. While boundaries are, strictly speaking, Euclidian lines with length but without breadth, frontiers have both length and breadth.

In ancient times, and even today in certain parts of the world which are either thinly peopled or not highly developed politically, it was common for different political areas to be separated from each other by zones rather than lines. Such zones were sometimes referred to as *marcher zones* and often lay in difficult terrain. It was in the mutual interest of the adjoining states that they should exist, and many of them persisted for centuries. However, with the growing scarcity of land and increasing cupidity on the part of men, it has become necessary to draw more precise lines in place of frontiers.

Since boundaries are essentially dividing lines, the extent to which they effectively divide is important. This is why obstacle boundaries provided by nature, such as rivers and high mountains, are so common and prominent. Although such boundaries follow natural features, they are not on that account natural boundaries, since the features along which they are drawn have the characteristics of frontiers rather than of boundaries. Indeed, from what has been said above, there can be no such thing as a natural boundary; logically, only natural frontiers are possible.

One of the most authoritative works on boundaries, *International Boundaries* by S. W. Boggs classifies boundaries into four main groups as follows:

(1) Physical boundaries, e.g. mountains, lakes, rivers, swamps, deserts.

(2) Geometrical boundaries, e.g. meridians or other great circles, lines of latitude and longitude or lines running parallel to them, arcs of circles, or any straight lines or lines drawn parallel to some natural feature which may itself be irregular, e.g. rivers and coastlines.

2 By frontiers Lord Curzon obviously meant boundaries which, as indicated in the paragraph following, are not synonymous with frontiers.

(3) Anthropogeographic boundaries, e.g. tribal boundaries, linguistic boundaries, religious boundaries, private property lines. Such boundaries are perhaps better described as cultural boundaries because they follow features that are man-made or which relate to human distributions.

(4) Complex boundaries, i.e. boundaries which consist of a combination of two or more of the above types.

States, such as the United Kingdom, which are wholly or almost wholly surrounded by sea, are fortunately exempt from many of the problems arising from international boundaries, although in the case of the United Kingdom the land boundary between Northern Ireland and Eire still presents some problems. But even the most completely insular state has still to reckon with the question of delimiting its offshore limits. And with the advent of the air age and the rocket age, new kinds of boundary problems have begun to rear their heads.

So far as the drawing of land boundaries is concerned, two vital and complementary processes are involved. The first is the settlement of the boundary on paper by the interested parties. This is the process of delimitation, while the second, the process of demarcation, is the actual identification and marking of the boundary on the ground.

The demarcation of boundaries often raises difficult problems because it is frequently found that the neat lines agreed upon at the delimitation stage are incapable of implementation in practice owing to the fact that very few maps are able to give an accurate and detailed representation of conditions on the ground. Consequently, boundary commissions charged with the task of demarcation must exercise a lot of discretion and imagination in order to arrive at boundaries that truly serve the purposes for which they were designed.

Until recently, the idea of defence and protection against attack was very strong and often uppermost in people's minds when fixing the boundaries of their states. But this attitude was in itself a cause of conflict since it invariably led to one state trying to secure advantages at the expense of its neighbour or neighbours. Thanks to the recent emergence of a world community of nations and the growing recognition accorded to international law, it is now much more difficult for any state to get away with boundaries which are advantageous solely to itself without in any way serving the interests of its neighbours. For this reason and also because of modern developments in offensive weapons and in means of transport and communications, especially the aeroplane, the idea of protective or defensive boundaries has been rendered obsolete.

Today, it would seem that those boundaries have the best chance of survival that confer no special advantages on any one party but have simply been agreed upon and drawn arbitrarily by the states concerned in recognition of the need for a line of separation between their respective domains.

The best examples of such boundaries are to be found in America, in particular the famous 49th parallel which forms the greater part of the boundary between the United States and Canada. But perhaps the most outstanding example in the world of two nations consciously resolving to settle their boundary disputes by peaceful means is that of Chile and Argentina, who agreed in 1898 to submit a long-standing boundary dispute to Queen Victoria for arbitration (see Boggs 1940). When the final decision was announced on 20 November 1902, the two countries signified their acceptance by erecting a giant statue of Christ – the Christ of the Andes – at the Uspallata Pass, bearing the inscription: 'Sooner shall these mountains crumble to dust than the people of Chile and Argentina shall violate the treaty which they have signed this day at the feet of Christ the Redeemer.'

RESIDENT POPULATION

As an element of political geography population is second in importance only to the territorial base, for just as without land there can be no question of a state, so also it is inconceivable to speak of a state in practical terms unless there are people who live within its territory.

Where the resource bases are comparable, a state with a large population has many advantages over one with a small population. Large populations were particularly advantageous in the days when the level of technology was low and national power was more or less directly measurable by the number of people capable of bearing arms. Nowadays when we talk about population in political geography we have in mind not only the sheer numbers but the effectiveness of the population in terms of its physical fitness, the skills it commands and the extent to which it is equipped to perform the various tasks which the modern state requires of its citizens in order to remain economically and politically viable.

In considering the size of a country's population the crudest and most common method employed is to look at the actual size of the population either alone or in relation to the total land area of the country, thus arriving at a ratio of persons to a given unit of area. This method permits the making of reasonably valid comparisons between conditions in different countries, but conveys the impression that all land has the same capability for supporting population. Since this is patently not the case it is much more helpful to consider population in an agricultural country, for instance, in relation to the area of available agricultural land, and in an industrial country, in relation to the industrial potential as measured by natural and human resources. Given a reasonably adequate economic base, it is definitely advantageous for a country to have a large population because this gives it greater geographical and political effectiveness.

Other factors, such as technical skills, political cohesiveness, loyalty and discipline can add significantly to the effectiveness of any population. Another factor of importance, especially from a military point of view, is morale, without which no people can be relied upon to resist outside attack or stand up for their rights.

Population is a dynamic phenomenon; it is changing all the time. Change reflects itself principally in the total numbers of people and in their sex and age composition. The principal cause of population growth is an excess of births over deaths, while population decline is brought about usually by the opposite state of affairs. But there are other factors like immigration, emigration and internal migration which can also affect population size in important ways.

The age composition of women is of particular importance demographically, and, since fertility is mainly confined to the age group 15–45, it is this group's size, fertility and mating tendencies that mainly determine the future growth of any population. If for some reason or other women of child-bearing age are either unable or unwilling to bear children, then the population is in danger of decline. This has often happened in situations where owing to such calamities as war there has been a dire shortage of men for reproductive purposes, especially in monogamous societies. But serious as such cases are, a worse state of affairs would occur if any calamity – natural or man-made – were to take a similar toll of the female population leaving a large surplus of men. And of course any disease that causes sterility in both sexes can have serious effects on the population, as for instance where venereal disease is rife.

Despite all these possible hazards, the general tendency in the world is for populations to grow. It is estimated that whereas in 1850 the total world population was 1,091 million, in 1940 the figure was 2,406 million, and in 1970 around 3,632 million. This growth has occurred in most cases without any special encouragement from any quarter; only in a few isolated cases in recent times have deliberate measures been taken by governments to encourage reproduction. More common have been attempts to reduce the population.

Abortion and the killing of children at birth or infanticide are known to have been quite common among primitive peoples whose land and other natural resources were limited and who therefore had to ensure that there were never at any time more mouths than could be fed. But there have also been societies, both simple and advanced, which have worked on the belief that human life is sacred and desirable and that it is their duty to have as many children as it is humanly possible to have.

In many of the developed countries where standards of living are high and the resource base seems infinite, we see the curious situation today where people plan deliberately to have small families. The explanation usually given for this apparent contradiction is that the rearing and education of children is an expensive process and unless the number of children is controlled parents will have to accept lower living standards for themselves and do less than they consider to be their duty by their children. It requires a considerable degree of sophistication to accept such a philosophy and it is little wonder that people in most of the rest of the world where living standards are much lower have not as yet been moved by such considerations. But some significant changes have occurred in recent years; India, which for many years resisted all attempts at birth control has recently accepted it as a matter of national policy and many other developing countries are beginning to follow suit.

The man who first gave currency to the idea that unchecked population growth was harmful for nations was Thomas Malthus (1766–1834). He wrote in 1798, in his *First Essay on Population*, that because of differences in their rates of growth, population tended to outstrip the means of subsistence unless it was checked by such calamities as disease and epidemics, wars and famine, which were far more effective in their operation than privately imposed checks like moral restraint and birth control. However that may be, we find that in most countries of the world the highest population growth tends to occur among the poorer sections, while in the world as a whole it is among the less-developed countries such as those of Asia, Africa and Latin America that the highest rates of growth occur.

The reasons for this are various. To start with the position within individual countries, it is generally the case that the wealthier classes are often more concerned about the maintenance of certain material standards of living than the poorer classes and are more prepared to adopt and often better able to afford the measures required to limit the size of their families than the poorer sections of the population. And the same tendencies in a greatly multiplied form manifest themselves as between the rich countries and the poor ones where the addiction to the material things of life is much less strong.

In normal circumstances population grows exponentially. Where a population is growing at an annual rate of increase of 1.5 per cent it can be expected to double itself every forty-seven years, while where the rate of increase is 2.8 per cent as was Ghana's in 1960 the population can be expected to double itself every twenty-five years. Considering recent advances in medical science and the greatly extended life expectancy in most countries, these figures suggest that we are going to witness tremendous population increases in the future.

However spectacular the size of a population may be, its quality is even more important, because in the final analysis this is the factor that determines the effectiveness of the population and its ability to maintain itself even in seemingly impossible physical circumstances. Among the many factors that determine the quality of a population may be mentioned three that are of particular importance in political geography. These are *technical efficiency, moral fibre* and *cohesiveness*.

The technical efficiency of a population is generally reflected in the level of literacy and in the quality of the educational system as a whole. The latter is much more important than mere literacy which by itself can add little or nothing to people's competence for the performance of essential tasks needed to maintain the state. Certainly, at a time when means of mass communication other than through the written word are so widespread, it is possible for many people who are illiterate or not particularly literate to be given quite a useful education. Especially is this true of the new or developing countries. The present world is a technological world and it is therefore important for people to have certain basic technical skills in order to maintain the basis for civilised existence. Whether people are engaged in farming or in manufacturing or in office work, they must be able to work efficiently by the thorough mastery of their work and the ability to utilise their intelligence and ingenuity to good advantage.

The moral fibre of a population is not easy to define or measure. Generally speaking, it has to do with the total philosophy and outlook of a people, in particular their ability to act resolutely and courageously on the basis of their convictions. If a nation wants peace and harmony then one should expect to see these qualities reflected in the people in the form of tolerance and consideration for other people. And, of course, one would expect to see a high premium set by the society concerned on these qualities in its individual members. On the other hand, where a nation believes in aggressiveness one would expect to see a high premium placed on aggressiveness and competitiveness in its citizens and a tendency to despise peaceful and reasonable conduct as being signs of weakness.

Many nations in the modern world are torn between these two extremes. While individual citizens are encouraged to be reasonable and peace-loving, the state itself exhibits aggressiveness in its dealings with other states, often amid loud proclamations of peaceful intention. There is thus in many countries a serious conflict between private and public morality. Within our own immediate circle of friends and acquaintances we are expected to be reasonable, considerate and altruistic, but when we face the world we are encouraged to urge our own claims as far as we possibly can, always putting our interests and convenience before those of others. Competition is the accepted order of the day and self-assertion has become the key to success. In spite of this there are certain qualities such as honesty, integrity, industry and courage that are tremendous assets, and any nation whose citizens are largely imbued with these qualities can ride many storms that might easily overwhelm those less well endowed with moral fibre.

Since a sense of unity is essential for the smooth functioning and the survival of any state, any characteristics or tendencies in the population which make for disunity call for careful attention. One of the most common of these is differences in language; another is differences in colour and other physical characteristics. By means of a conscious process of education linguistic differences can be reduced or even eliminated within a matter of one or two generations, as has happened in the United States and is currently happening in Israel; but it is much more difficult to eliminate differences in colour.

In his 1965 Reith Lectures on the theme *A World of Peoples* Robert Gardiner of Ghana made an analysis of this question in great depth. Colour creates differences not just in itself but also because of the many curious and often mistaken associations that go with it. In spite of all the available evidence and the arguments that have been advanced to the contrary, the belief still persists in many circles and in many parts of the world that people of certain colours, especially black or brown, are inherently inferior to those of certain other colours, especially white.

It is usually the case that people occupying one territorial area belong to more or less the same colour or racial stock, but there are a number of instances in the world where owing to the accidents of history this is not so. It is in such areas that colour problems come to a head, especially where the numbers composing the different groups are sufficiently large to pose the threat of economic or political competition or of possible dominance by one group over the rest. The best examples in Africa of countries placed in this situation are to be found in the southern part of the continent, especially the Republic of South Africa and Rhodesia.

Racial problems can create many difficulties in the countries where they occur. But while in some of these countries, such as the United States, Britain and certain countries in the West Indies and Latin America the state either does not officially support racial prejudice and discrimination or is making serious efforts to reduce or eliminate them altogether, in South Africa the entire apparatus of the state is employed to perpetuate racial discrimination, which has been raised to the level of an official national policy, namely, apartheid, based on the myth that the white races are superior to the other racial groups to be found in the country. There are other countries where even though racial discrimination as such is not practised or supported by the state, no conscious attempt is made at integration. Here, society is structured along parallel lines, and although it is recognised that the different racial or colour groups are economically interdependent, yet socially there is little or no intermingling among them. J. S. Furnivall many years ago described such societies as 'plural societies' (Furnivall 1939); but nowadays the term most commonly employed for them is 'multiracial societies'.

Rather more subtle than colour differences but equally harmful and insidious in their effects are differences in tribe, which are denoted by the term 'tribalism'. Tribalism has proved to be the bane of many of the new nations of Africa. It expresses itself in differences in language or dialect, of customs and attitudes. Like colour and racial differences, tribalism is often perpetuated by the notion of stereotypes, that is, the notion of imaginary attributes, often uncomplimentary in nature, which are supposed to characterise persons belonging to groups other than one's own in order to justify hatred or contempt for them. The tendency for people who share certain common attributes, be they in the form of language, cultural or physical characteristics, to feel a sense of identity is something that cannot be banished completely from the minds of most human beings. What is important is that such differences should not be exaggerated to the point of placing people in hostile camps even where they are supposed to share a common citizenship. Indeed, if the various stereotypes that are current in any given society can be seen as a joke, they are capable of providing an important basis for harmonious coexistence between different groupings within the society in question by developing in them the ability to laugh at their own weaknesses and to admire the positive qualities of other groupings.

Quite obviously, any society which is faced with the problem of racial or tribal prejudice is placed at a serious disadvantage in any attempt it may wish to make at mobilising the whole of its available human resources for the performance of those national tasks which demand internal harmony and unity as a precondition for success.

SYSTEMS OF GOVERNMENT

Every politically organised area requires a system of government or an administrative structure; this is inherent in the very idea of political organisation and there is no choice in the matter short of complete anarchy. What can and does vary from area to area is the character of the system of government in force and its degree of complexity and efficiency. Nowhere is this more clearly and strikingly demonstrated than at international boundaries, where, despite any general similarity which may exist in physical conditions on either side of the boundary, one often observes striking cultural differences due largely to the effects of different systems of government. Because of the differences in the legacies bequeathed by the various colonial powers Africa abounds in examples of this kind, and a road journey from Ghana to Nigeria through Togo and Benin offers an interesting object lesson in the potency of what Derwent Whittlesey has described as 'the pattern of culture' (Whittlesey 1938).

Closely related to the question of administrative structure is that of political status. Politically organised areas vary in status from the wholly sovereign to those that are wholly subject to other areas. Sovereignty has many meanings and implies a whole range of attributes, but in its simplest sense, as applied to political entities, it means the right to declare war or make peace without reference to any other power. In the past such absolute sovereignty has been possible both in theory as well as in practice. Today, however, thanks to the greater ease of movement and the marked degree of interdependence which has come to exist between different countries and states, it is virtually impossible in practice for any state, however powerful it may be, to exercise absolute sovereignty and independence in its actions, except in relatively unimportant matters which do not significantly affect the interests of other states. But for small and weak states the exercise of sovereignty, except in purely domestic matters, is almost entirely out of the question in the modern world. Nevertheless, sovereignty is a factor of importance in discussing the systems of government in operation in politically organised areas.

The first general distinction to be made in discussing systems of government is that between sovereign and non-sovereign states. If it is sovereign then it has the right to manage its domestic affairs without reference to any other power.

Non-sovereign states can vary very much in the degree of autonomy which they enjoy and in the extent of their subordination to other states or powers. Taking the two opposite ends of the scale, non-sovereign states may be divided into two principal types: those that are wholly dependent in both external and internal matters and those that enjoy autonomy in respect of their internal matters. The former are colonial possessions in the traditional sense, while the latter are self-governing territories. In almost all the former British colonies in Africa and elsewhere, which are now independent, internal self-government has marked the penultimate stage in the process of evolution towards complete independence.

Since wholly dependent countries have no say either in their internal affairs or in their external affairs, it can be said that the will of the people finds little or no expression in government. But a note of caution must be sounded here. There are very few countries, however dependent they may be politically, where the people have no say at all in the running of affairs even at the lowest level. What we really mean when we say that the will of the people finds little or no expression in government is that the control of the central government, though not necessarily of local government, is not effectively in their hands. In the same way, when we talk about countries which have self-government we mean those countries that are fully in control of their internal affairs though subject to a foreign power in respect of their external relations. The Indian princely states, such as Hyderabad and Mysore, were in this position during the period of British rule. As Reginald Coupland has observed, they were broadly speaking autonomous as regards their domestic affairs, but accepted the suzerainty of the Crown and its control of their external relations (Coupland 1943).

Until the First World War of 1914–18 sovereign countries had control only of colonial possessions, and this control was directly exercised. After the war, when the League of Nations was formed, the government of the former colonies of Germany and of certain parts of the Turkish Empire was transferred to some of the victorious powers as Mandates under Article 22 of the Covenant of the League of Nations. Although the powers concerned had direct control over these territories, they administered them on behalf of the League through its Mandates Commission.

Three classes of Mandates (A, B and C) were recognised. Class A Mandates were the former Turkish communities of Syria and the Lebanon, Iraq, Palestine and Transjordan. They were provisionally recognised as independent nations but with the proviso that they must accept, for a time, advice and assistance from more advanced powers. Class B Mandates were mostly to be found in tropical Africa and comprised Ruanda and Urundi, Tanganyika, French Cameroons, British Cameroons, French Togoland, British Togoland. They were to be administered by mandatory powers able to guarantee freedom and public order and equal opportunities for trade under specially devised governments and regulations. Class C Mandates consisted of South West Africa and a number of sparsely settled Pacific islands. They were to be administered according to the laws of the mandatory power as an integral part of its territory, subject to safeguards in the interests of the population. Apart from South West Africa the territories within this group comprised Yap and other former German North Pacific islands, German New Guinea and certain adjacent South Pacific islands, Nauru, and Western Samoa (Bowman 1928).

It was explained at the time of the creation of these Mandates that they needed tutelage because they could not stand alone under the strenuous conditions of the modern world. Most of them were still not independent at the time of the Second World War of 1939–45, and after the war they were renamed Trust Territories under the United Nations. Considering the mood of the period following the Second World War, the designation 'Trust Territories' was more appropriate than the previous one because it emphasised the fact that the powers in direct charge of these territories were simply governing them *in trust* for the inhabitants until they should be in a position to administer their own affairs.

As has been pointed out already, however, all dependent territories, including Trust Territories, are basically colonies. On etymological grounds, our current use of the word 'colony' is not entirely correct. In the original Greek sense, a colony was an independent city founded by emigrants, and several such colonies were established by Greeks in the Mediterranean basin. Later, when Rome became a great power, she also established colonies in various parts of her empire within and outside the Mediterranean basin. Unlike the Greek ones, the Roman colonies consisted of settlements made up usually of veterans in conquered territory acting as a garrison. When, many centuries later, Britain and other European nations began to establish colonies in America and elsewhere, they were essentially settlements in the Greek sense, inasmuch as they represented transplantations of people from the mother country to distant places.

But apart from these genuine colonies, many European nations established themselves in a number of tropical countries all around the world not with a view to settlement but with a view to trade and economic exploitation. Quite contrary to the original meaning of the term, such outposts of empire were also described as colonies, and indeed it is in this sense that the word 'colony' is now most commonly used and understood.

Altogether, three different types of colonies are now generally distinguished: colonies of settlement, colonies of exploitation, and strategic colonies. In colonies of settlement the population consists predominantly of people from the mother country, while in colonies of exploitation the bulk of the population are indigenous to the colony itself, although political control is in the hands of aliens. In such circumstances it is hardly appropriate to describe the controlling power as the mother country, and the term 'metropolitan country' which has gained currency in recent years is much more suitable. Strategic colonies are strictly speaking places like Gibraltar, Aden, Hong Kong and Guam, whose principal value to the metropolitan country in control lies in their strategic location.

Despite the above distinctions, it is possible in practice for a single area to combine the characteristics of more than one of these three types of colonies. Certainly, the strategic function is frequently in evidence in the two remaining types, since there are many countries which though more extensive than mere bases are strategically located in terms of regional or global defence. A good example of a country which is all three types of colony at once is South Africa. Here a resident white population from Europe is itself operating like a colonial power in respect of the land and its African population and treating both as objects of economic exploitation. Finally, South Africa's location at the southern tip of the African continent gives it an important strategic advantage.

We come next to the actual systems of government which can operate in politically organised areas. In order to simplify the discussion, we shall confine ourselves to the sovereign state and leave the reader himself to apply any relevant conclusions which emerge to other types of politically organised areas.

The test of a government's effectiveness is essentially a pragmatic affair, but first and foremost it must have control over its territory. In other words, it must be in a position to exact obedience in accordance with the norms of the people concerned. In the Greek city state of twenty-five centuries ago, government was within easy reach of every part of the state and in close touch with all the citizenry. In a sense, this is the situation which prevails in certain of the Swiss cantons even today. But as states grow in size and population, this intimate personal contact becomes increasingly difficult to achieve, and more elaborate and less personal systems of control are demanded. The resultant forms of government are mostly what we see in the world today, and they may take one of two principal forms: the unitary state, on the one hand, and the federal state, on the other.

The relative merits of these two forms remain the subject of endless debate, especially in present-day Africa, where there is a desperate search for effective and stable political models. On the whole, federal government is better suited to large and greatly diversified states than to small and generally homogeneous ones. But even in the smallest state loyalty can never be monolithic, and a certain measure of local autonomy and governmental delegation is necessary in the sheer interest of efficiency and contentment.

In every state there are two opposing sets of forces: centripetal and centrifugal. The centripetal forces consist of those factors, physical and psychological, which tend to bring the people of the state together; and they usually relate to things or features which the people either have, or think they have, in common – language, customs and mode of life, religion, history, skin colour or other bodily characteristics and even diet. But these same things, if they are not congruent, can exert strong centrifugal or divisive influences. Centrifugal forces are especially strong in a state if they can find nearby centres, often in neighbouring states, to which they are attracted and upon which new loyalties can focus.

It is important for governments to recognise these different forces, to foster those that are centripetal and diminish the effects of those that are centrifugal. Although the physical factors that promote unity are quite important, the psychological ones can be even more potent. And one of these is the cultivation and acceptance of a national myth calculated to foster the idea that there is something special about the citizens of the state in question and that they have more or less been marked out by Providence to fulfil a special mission in the world.

The deliberate fostering of unity in most states, especially in the new states of Africa, is necessary because the majority of these states are no more than artificial creations or geographical expressions. Unlike the family, which is a natural biological unit, states are purely human creations. Consequently, in all states the idea of unity must be held constantly before the people; and there is no better way of doing this than by inducing them to believe that they have always been united or that they have suffered together in the past and therefore should hold together in the present and in the future in order to be able to resist encroachments from outside. It is also useful for them to believe that they have a common and great history, which it is their duty to uphold, and an even greater destiny which it is their duty to fulfil, although clearly it is not desirable for any nation to lay too much stress on these negative attributes. That is why Julian Huxley defined a nation somewhat cynically as 'a society united by a common error as to its origin and a common aversion to its neighbours' (Huxley 1939).

Special problems of government in African countries

Right up until the Second World War all the countries of Africa, with the exception of Egypt and Liberia, were under European colonial rule, with the indigenous populations taking little or no real share in the serious processes of government. After the war the African independence movement began to gather momentum: Ethiopia regained its independence from Italy in 1941, and other African countries gradually began to gain self-government from their former rulers under a variety of democratic constitutions.

In the majority of cases these constitutions were based on a unitary form of government; but in a few exceptional cases, as in Nigeria, a federal form of government was adopted. In all cases, however, it was accepted or assumed that the operation of the constitution would be based on a multi-party form of government in accordance with the principles of western parliamentary democracy. Most of these expectations proved to be illusory. In one country after another the multi-party form of government was either replaced by a single-party system or the whole idea of parliamentary democracy was abandoned and replaced by a presidential form of government dominated by a single personality. Alternatively, the original constitutions were abolished and civilian government gave way to military rule with or without civilian participation. Thus, with a very few exceptions, the resulting picture in Africa since the achievement of independence has been one of widespread political instability with all its attendant social and economic consequences.

In the face of all these changes it is difficult to foresee what the eventual political pattern will be in Africa. So vast are the economic and social problems facing most of the new nations in the continent that there is little prospect of purposeful development unless strong government can be established, capable of pushing through much-needed programmes of reform and overcoming the disruptive and divisive forces of tribalism and regional particularism. Civilian government has proved ineffective in many cases, but it is not certain whether military rule will work any better or provide a greater degree of stability. The answer might well be a blend of the two, at least for the time being. Whatever the outcome, one thing seems certain beyond all doubt. It is absolutely essential that the economic foundations of the new African nations should be greatly strengthened as quickly as possible so that the ordinary man and woman can be guaranteed a reasonable standard of living. Unless this objective is achieved the prospects for political stability will continue to remain very slender.

ECONOMIC BASE

By the term *economic base* is meant all those economic resources and activities that provide material sustenance for the state and its population as well as the manner in which they are organised. There are important and obvious links between the economic structure, the population and the general political life of the state. Other things being equal, a healthy, well-educated and contented population makes for a vigorous economy, while a healthy economy conduces to the well-being of the population. And if both the economy and the population are in a sound state the political life of the state stands a much better chance of proceeding smoothly and effectively.

The first foundation of a sound economy is the available resource base; for a country's wealth and standard of living depend to a large extent on the extent, range and quality of its natural and human resources. Although the population of any given country does not have to depend entirely on the resources of its particular area, it must have the capability to pay out of its own resources for whatever goods or services are obtained from outside. The more complex and advanced a society is, the greater, generally, are its resource demands.

Some resources, such as land, that is, the soil, the vegetation, water bodies, fauna and mineral wealth, are tangible and material, while others, such as human skill and ingenuity, are intangible and are present only in the form of a potential. Human ingenuity is one of the most important resources that any country can have, and over the years it has come to play an increasingly decisive role in determining the real extent of a country's wealth. Without human ingenuity natural resources are largely inert, hence the truth of Carl Sauer's observation that 'natural resources are in fact human appraisals' (Sauer 1952).

Since food is man's prime need, the area of cultivable land or of exploitable stretches of water constitutes the most important material resource available to any society. Where a country does not possess any agricultural land or enough of it to meet its basic requirements it is essential that it should have other resources or forms of wealth to compensate for the deficiency, usually in the form of minerals. A good example of this is provided by some of the Middle East countries situated on the Arabian peninsula, which consist almost entirely of desert without enough water to make even irrigated agriculture possible on any significant scale. Some of these countries are able to maintain a high degree of economic prosperity because of their wealth in petroleum. Without this asset human life would be extremely precarious.

Economic activity occurs in varying forms and at varying levels. Where man is concerned simply with cultivation, fishing, forestry and mining, the resulting activity is described as primary. On the other hand, where he is concerned with the processing of the products yielded by primary activities or other forms of manufacturing he is said to be engaged in secondary activities.

Within the area of primary production there is an important distinction between the exploitation of what may be termed broadly as biological resources and that of mineral resources. Given proper methods of husbandry, biological resources may be exploited almost indefinitely through a combination of human action and natural processes of growth. Indeed, with proper management such resources can be greatly increased in quantity and quality. In contrast, minerals are finite resources; what is available cannot be added to, and what is exploited is permanently lost.

It is very rarely that the economy of a country is characterised either wholly by primary production or by secondary production; usually the two are combined in varying proportions, although the tendency is for secondary activities to assume an increasing importance as a country advances in its development and level of prosperity.

All manufacturing involves the use of two ingredients in addition to the basic raw material or materials. These are human ingenuity, on the one hand, and some kind of power or energy supply, whether it be derived from human brawn or animal power or from power supplied directly or indirectly by water, fossil fuels, such as coal or petroleum, or by tides, winds, solar energy or nuclear energy. Nowadays, however, when we talk of 'manufacturing' we have in mind large-scale operations involving the use of highly advanced techniques and the investment of large amounts of capital. Thus, in contrast to simple economies, which are essentially primary or in which manufacturing takes place only in a very rudimentary sense, we have to recognise complex economies or industrial societies in which even the primary activities which occur are highly mechanised, highly capitalised, and highly organised.

It goes without saying that the political and other institutional arrangements which are required to sustain a highly industrialised society are far more complex and far more delicately balanced than those required for maintaining a society that is essentially primary. Industrialised societies are far more productive than primary ones, but in many ways they are also far more vulnerable to dislocation. Let one link in the production chain get out of order and trouble immediately ensues. In other words, precisely because a highly industrialised society is also a highly specialised one, the degree of interdependence between its various sectors is very much greater. Moreover, the infrastructure required to maintain such a society often extends far beyond its own immediate borders. Such societies must therefore perforce concern themselves with international relationships – not just power relationships, but relationships in the fields of international trade, international finance, and international transport and communications.

The proven material advantages which industrialised societies have over those engaged predominantly in primary production have led to an intense desire among many of the developing nations for increased industrialisation. Unfortunately, however, because of their failure to consider properly the implications of industrialisation before making the venture, many of them have run into serious difficulties and saddled their fragile economies with huge foreign debts.

The possible bases of successful industrialisation are many. They include in particular an adequate supply of: (1) raw materials, (2) power, (3) labour with requisite skills, (4) capital, (5) people with entrepreneurial skills, and (6) markets, both internal and external. It is not necessary to embark on a detailed discussion of all the above bases of industrialisation. The ones that particularly deserve attention because of their special relevance to political geography are raw materials and power, especially petroleum, whose availability (whether from local or outside sources) has now become an indispensable prerequisite for modern economic existence.

Because of the importance of raw materials and power in industry, most of the more highly industrialised nations adopt various measures for ensuring that they have access to a fair range of essential raw materials, especially those that have strategic importance, as well as to petroleum and other sources of power. If they do not occur within their own boundaries they seek other means, such as trade and treaty arrangements, to guarantee their supply or the supply of acceptable substitutes. Another device adopted by some countries is to stockpile supplies of their most important raw materials, power and food requirements at home against unforeseen contingencies, such as war or economic hold-ups. But such expedients are at best only short term. In the long run the best and most reliable method available to all countries for securing the supply of their essential requirements is through international trade. Generally speaking, however, the more self-sufficient a country can be in industrial raw materials, power and food products, the greater are its chances of political and military power.

Even so, it must be recognised that the present economic and military power of the various countries of the world has been influenced to a very great extent by historical accidents. Many of the advanced and highly industrialised countries have achieved their present position not so much because of their intrinsic wealth in natural resources as because of their political power and their ability to obtain cheap and ready supplies of raw materials from the less developed parts of the world, notably the tropics. Up till the end of the Second World War most of these tropical areas were colonial possessions controlled by the great powers, and their economies were characterised by the export of raw materials to the more advanced countries in return for manufactured consumer goods. This pattern is now undergoing fundamental changes. Most of the primary producing countries are now politically independent and are seeking to raise their living standards through increased industrialisation. This means that they are likely to export less of their products in the form of unprocessed raw materials and to import rather smaller quantities of manufactured goods from the industrialised nations. What is even more important, because they are now politically independent they are free, at least in theory, to choose their trading partners, and the flow of vital raw materials from them to particular industrialised countries can no longer be taken for granted. This is one of the main reasons why the political orientation of many of the developing nations has become so important in present-day international politics. Instead of trade following the flag, as was the position only a few decades ago, today the tendency is for trade to follow political ideology. Nevertheless, historical associations die hard, and even though significant changes are taking place in the actual composition of exports and imports, most of the developing countries that have recently achieved political independence still conduct most of their trade with their former colonial rulers.

No discussion of economic resources can be complete without reference to human resources. As has already been stated, natural resources are in themselves largely inert and become effective only when human ingenuity operates upon them. Culture and technology are vitally important in resource development, and provided they are available and able to exert their due influence there is hardly anything in nature that cannot be put to some use for the benefit of man. This is why education is such an important aspect of economic development.

Two other important human factors are the availability of labour and entrepreneurial skills. Effective labour implies, among other things, the availability of the right skills and the right social and political orientation. Again, education is able to impart many of these qualities. The imparting of entrepreneurial skills is another matter; it depends much more on individual personality and character. Even so, the type of education and upbringing a person receives and the general outlook of the society within which he is brought up and lives can go far towards determining whether or not he will develop the qualities of daring and independence demanded of entrepreneurs.

Finally, we come to capital. This is an area where most of the new African nations are at their weakest because their heavy dependence in the past on primary production has given them very little opportunity for the accumulation of surplus capital. The tendency has therefore been for them to depend on capital from the more highly developed nations, whose capital has been principally responsible over a very long period for most of the economic and social developments that have taken place in the rest of the world. Even the development of the raw materials in most of the less highly industrialised parts of the world has come about mainly through the operation of capital from the industrialised nations in the creation of roads and railways and trade arteries and in the provision of new incentives for production.

Arising out of all this is the obvious though unpalatable truth that unless the newly independent African nations find the means to provide all or most of the capital required for their economic development they can never hope to achieve economic independence or enjoy the full material benefits of their recently acquired political independence.

TRANSPORT AND COMMUNICATIONS

The words *transport* and *communications*, though closely related in meaning, are by no means synonymous. Transport refers to the physical movement or carriage of goods and people and the various means employed for the purpose, while the word communications refers more specifically to the transmission of ideas and messages as well as the physical connections which exist between places for the purpose of transporting or exchanging goods, people, ideas and information. Thus, all the known forms of transport can also be employed for establishing communications between different places and people, although not all means of communication are suitable for purposes of transportation.

The importance of transport and communications in any politically organised area is too obvious to require much further elaboration. Without an effective system of transport and communications no modern state can function properly either politically or economically. The larger the area covered by a state the greater is the need for its various parts to be in effective contact with each other; similarly, the more developed a country is the more complex is its system of transport and communications likely to be. Lines of transport and communications usually form a network within any given area, with certain places standing out as focal points. Because of the dominant role played by capital cities and towns as centres of political and, often, economic activity, they are usually the most prominent focal points in the system of transport and communications within most states. Below the capital come other centres in descending order of importance, each serving as a focal point for its particular area.

Individual states and different parts of the same state differ as regards the physical advantages or disadvantages they possess for the development and use of the various forms of transport and communications. Some of the most serious problems in this respect are presented by mountainous terrain, deserts, thick tropical and equatorial forests, swamps and marshes. At one time such obstacles made transport and communications virtually impossible. Today, however, thanks to modern advances in science and technology, there are hardly any physical obstacles that cannot somehow be overcome by man, although the price in terms of money, time and effort may be very considerable.

The principal forms of transport in use are human transport, animal transport, water transport, rail transport, road or motor transport, and, finally, air transport. Human transport is the most primitive, but it is still the most universal and in many of the less developed parts of the world often the most important. Animal transport is very old and almost as primitive, but its importance is fast diminishing. Water transport too has a long history, but has undergone such remarkable changes and improvements over the centuries that its more developed forms are now counted along with road, rail and air transport among the recognised forms of modern transport.

Apart from their use for the conveyance of goods and people all the various forms of transport mentioned above have military and strategic uses. Especially important in this respect is air transport, which has

34

completely revolutionised the entire concept of war and strategy in the modern world, rendering obsolete many of the ideas formerly held about security and the protection afforded by natural barriers. In spite of all this, the military and strategic importance of ships still remains considerable, and the possession of a strong navy by any state constitutes a major military asset for offensive and defensive purposes.

The various forms of communication employed by man have also undergone remarkable changes and improvements through the centuries, and most modern states are able to employ a wide range of means for their internal and external communications, such as telegraphy, the telephone, radio or wireless and, most recently of all, television. For most ordinary people the usual forms of communication are still the written and the spoken message, which have necessarily to depend on one or other of the forms of transport already discussed and are therefore subject to their respective disadvantages. However, all the barriers inherent in this form of communication have been swept aside by radio and television which require no continuous physical links, and it is now possible for man by using these new media to communicate instantly with practically any part of the earth and even beyond to outer space, as has been amply demonstrated by recent manned and unmanned explorations into space. But even here, we are only at the beginning, and there is no doubt that the next few years will usher in many more spectacular developments which will knit individual states and the whole world even more closely together than is now the case. Exciting as these developments are, they nevertheless pose serious and even terrifying problems for mankind, inasmuch as the ability to communicate readily with people also implies the ability to exercise control over them from a single remote centre. Such power can offer many important benefits to man, but in the hands of an unscrupulous government it can also have devastating effects upon the freedom of the individual.

PART II

Introduction to the political geography of Africa

3

Africa: general introduction

Africa is a continent of vast proportions. It covers an area of 30,328,000 sq. km (11,700,000 sq. miles), which makes it the second largest continent in size after Asia, and contains a population of over 400,000,000, which is exceeded only by those of Asia and Europe. It is, however, the most compact of all the continents in terms of shape, measuring approximately 8,050 km (5,000 miles) from north to south as well as from east to west and being bounded by a coastline which is generally straight and relatively short in relation to the area of the continent. Another of Africa's distinguishing characteristics is the fact that the Equator runs almost exactly through its middle, giving it a markedly symmetrical arrangement of climatic and vegetational conditions in both its northern and southern halves ranging from equatorial and tropical forests through tropical savannas and deserts to Mediterranean conditions (Fig. 2). However, because of the much greater east–west extent of the northern half most of the climatic and vegetation belts cover a much larger area than their southern counterparts, as is well illustrated in the case of the Sahara desert, which covers some 10.4 million sq. km (4 million sq. miles) and is the largest desert of its kind in the world, with a north–south width of 1,800 km (1,100 miles) and an east–west extent of 5,600 km (3,500 miles).

In its structure and relief, also, Africa differs markedly from the other continents. Most of the surface is covered by very ancient crystalline, metamorphic and sedimentary rocks of great hardness, known collectively as the basement complex, with only limited areas of young sedimentary rocks such as are quite common in other parts of the world. The relief is also marked by extensive areas of plateaux which have been faulted, uplifted and tilted in many places. Young folded mountains are almost entirely absent, the only important examples being the Atlas ranges of North Africa, which are almost contemporaneous with the Alps of Europe, and the so-called folded ranges of the Cape region of South Africa which belong to a slightly older geological period. Elsewhere, most of the prominent relief is either the result of differential erosion or of fairly recent volcanic activity, as in the case of most of the East African mountains like Kilimanjaro (5,899 m or 19,340 ft), Kenya

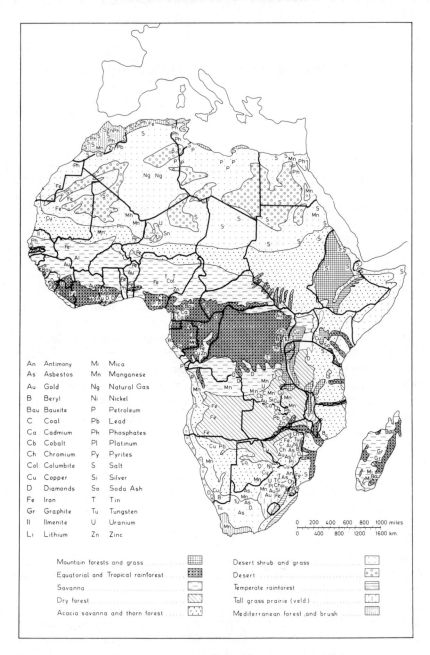

Fig. 2. Africa: vegetation and mineral resources

(5,203 m or 17,078 ft), Meru (4,569 m or 14,979 ft) and Elgon (4,053 m or 14,178 ft) and of the Cameroon Mountain in West Africa (4,073 m or 13,354 ft). Generally speaking, the highest areas in the continent, with average elevations of between 900 and 120 m (3,000–4,000 ft), are found in the southern and eastern sections approximately south of a line between the mouth of the Congo (Zaire) River and the Ethiopian massif, while most of the area to the north is characterised by considerably lower average altitudes (Fig. 3).

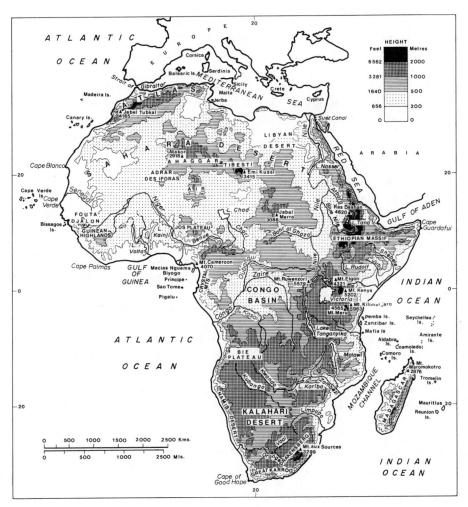

Fig. 3. Africa: physical features

Africa is not blessed with very many large rivers in proportion to its size, but it nevertheless contains some of the largest and longest rivers in the world, such as the Nile, the Congo, the Niger and the Zambezi. Most of these rivers are not very useful for transport and communication because their courses are frequently interrupted by rapids, especially where they descend the edge of the interior plateaux within a few miles of the coast. A good example is provided by the Congo which is interrupted by no fewer than thirty-two rapids and cataracts forming the Livingstone falls in the course of a descent 214 m (700 ft) over a stretch of 353 km (220 miles) just below Stanley pool as the river crosses the Crystal Mountains to make its way to the sea. Whereas in other parts of the world rivers have usually provided the easiest routes of entry inland from the sea, in Africa the rivers have more often impeded entry from outside for the reasons already given, coupled with the general absence of navigable estuaries at the mouths of many of the rivers and the marked seasonal fluctuations which they suffer in volume. Despite these deficiencies many of the rivers lend themselves readily for use in the production of hydro-electric power, thus giving Africa the highest potential among all the continents for the production of this form of energy whose exploitation has only just begun. It is estimated that out of the world's hydro-electric power potential of 555,000 megawatts Africa alone has 145,000 megawatts, i.e. some 26 per cent. However, the amount of hydro-electric power actually in harness in the continent in 1972 was only about 5.3 per cent of the total potential. The countries with the highest potential include Zaire (78,000 megawatts), Madagascar (11,500 megawatts), and Zambia (17,000 megawatts) (see p. 257).

Although it is customary to think of Africa as consisting of a single continental mass, in fact it includes a large number of offshore islands of varying sizes and geological origins. Of these the ones that can truly be said to form part of the continent because of their structural and political affinities are Madagascar (Malagasy Republic), the largest, Zanzibar and Pemba, the Comoros, Mauritius, Réunion and the Seychelles, all in the Indian Ocean, and the Cape Verde Islands, Fernando Po, Principe, São Tomé and Annobon, all in the Atlantic Ocean.

For purposes of geographical study Africa may be divided into the following seven parts or *regions*, using the word *region* in a somewhat loose sense (Fig. 4):

(1) *North Africa*, comprising Algeria, Egypt, Libya, Morocco, Tunisia, and the small Spanish possessions of Ceuta and Melilla.
(2) *West Africa*, comprising Benin (formerly Dahomey), the Cape Verde Islands, The Gambia, Ghana, Guinea, Guinea Bissau, Ivory Coast, Liberia, Mali, Mauritania, Niger, Nigeria,

Senegal, Sierra Leone, Western Sahara (formerly Spanish Sahara), Togo, and Upper Volta.

(3) *North-East Africa*, comprising Ethiopia, the Territory of the Afars and Issas, Somalia, and Sudan.

(4) *Central Africa*, comprising Burundi, Cameroon, the Central African Republic (renamed Central African Empire in December 1976), Chad, the Congo, Equatorial Guinea (consisting of Rio Muni, Fernando Po and Annobon), Gabon, Rwanda, São Tomé and Principe, and Zaire.

(5) *East Africa*, comprising Kenya, Tanzania (including the Islands of Zanzibar and Pemba), and Uganda.

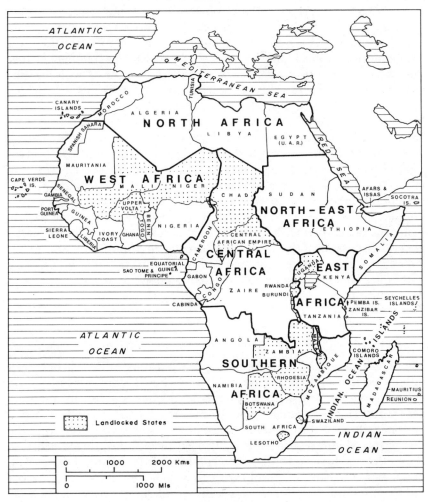

Fig. 4. Africa: political and regional divisions, showing landlocked states

43

(6) *Southern Africa*, comprising Angola, Botswana, Lesotho, Malawi, Mozambique, Namibia, Rhodesia (Zimbabwe), South Africa, Swaziland and Zambia.

(7) *Islands of the Indian Ocean*, comprising the Comoros, Madagascar (Malagasy Republic), Mauritius, Réunion and the Seychelles.

Altogether, therefore, there are fifty-four countries in Africa, including the islands of the Atlantic and Indian Oceans, of which all except South Africa, Namibia, Rhodesia, Western Sahara, and Réunion were independent and under African control at the beginning of 1978. In the case of the first three the governments in power are white minority governments, while in the case of Western Sahara the position was that even though the former colonial power (Spain) had withdrawn, the territory had been forcibly partitioned by two of the neighbouring states, namely, Morocco and Mauritania. Réunion's position was rather more complex in that the country formed a department of France and thus was not a sovereign state in its own right. To discuss the political geography of each of these many political units into which Africa is divided, including the offshore islands, would be an impossible and pointless task since many of the units concerned are comparatively small in size and population and share several common geographical and political characteristics with their neighbours. A much more practical approach is to consider them within their respective regional groupings, singling out any particular units which demand special consideration on account of their peculiar problems and characteristics. It is important to remember, however, that what we are primarily concerned with is the political geography of Africa and not a general discussion of the physical or human geography of the continent. This calls for a selective approach and a concentration on those aspects and topics which have a pertinent bearing upon the problems of political geography.

LOCATION, ENVIRONMENT AND HUMAN INTERACTIONS

There is little doubt that the political geography of Africa has been influenced in a number of important ways by the continent's location on the globe in relation to other land masses and by its physical geography (Fig. 5). Since ancient times the northern parts of Africa have had close cultural and political ties with southern Europe and the rest of the Mediterranean basin. The Strait of Gibraltar, which separates the north-western tip of the continent from Spain, measures no more than 12.5 km (7.75 miles) in the narrowest part and 38.2 km (23.75 miles) in the widest part, while until the opening of the Suez Canal in 1869 Africa had a continuous land connection with Asia across

44

Fig. 5. Africa in its global relationships

the narrow Isthmus of Suez. It was because of the proximity between North Africa and the Iberian peninsula that the Carthaginians were able to establish themselves in Spain and to use it as a base for attempting the overland conquest of Rome. Later, in the Middle Ages, the Moors were also able to cross over from Africa into Spain in the eighth century and to dominate it for the next seven centuries. Earlier, when the Greek Empire was at its height, a number of Greek colonies had been established on both the European and African shores of the Mediterranean, and for many centuries subsequently the greater part of North Africa, including Egypt, formed part of the Roman Empire as far south as the Saharan border. Even today, from a cultural point of view this part of Africa belongs more to the Mediterranean world than it does to the rest of the continent south of the Sahara, although in recent years the growing awareness of the importance of closer identity

45

with the newly independent states of Black Africa which dominate the Organisation for African Unity has begun to swing its outlook increasingly southwards.

In contrast with the long and close intercourse which North Africa has enjoyed with the lands around the Mediterranean and beyond, the history of Africa south of the Sahara has been marked, until comparatively recent times, by virtually complete isolation from the rest of the world. It is true that for some centuries before the great European voyages of discovery in the fifteenth century the northern fringes of the Sudan lands of West Africa were in contact with North Africa and the Middle East through the caravan routes across the Sahara and that the lands within and around the Horn of Africa also had links with the Arabian peninsula and possibly India; but for the most part Africa south of the Sahara formed a world of its own, remote and isolated (Fig. 5). The cause was not the Sahara alone; the vast distances separating this part of Africa from the other continents which lie widely scattered in the oceans of the Southern Hemisphere were equally responsible. Thus the distinction commonly made between North Africa and Africa south of the Sahara is not just one of convenience; it is firmly rooted in the facts of geography and history.

There is abundant historical evidence that Africa south of the Sahara was vaguely known to the peoples of Europe even before Roman times. It is recorded, for example, that round about 600 BC King Necho of Egypt sent a Phoenician expedition on a sea voyage around Africa, but this expedition appears to have made little physical contact with the Black inhabitants of this part of the continent. It was not until the middle of the fifteenth century that the first proper contacts were established between Black Africa and Europe as a result of a number of voyages undertaken by the Portuguese and Spanish who, following the blockage by the Arabs and Turks of the traditional trading routes to Asia through the eastern Mediterranean, were obliged to seek an alternative route around Africa. Several footholds in the form of forts were established along the coast of West Africa, especially in the area now known as Ghana, which served as trading posts between Africa and Europe, but the principal objective of these voyages was the rich lands of the Orient and the really important role of Africa was to provide convenient stopping points and victualling stations along the sea route to Asia.

However, after the discovery of America and the need which this created for cheap labour for work on the plantations, Africa assumed a new importance as a source of slaves, and the Guinea coast suddenly became the scene of intense activity involving practically all the maritime powers of Western Europe. Slaves were shipped from West Africa to America, from where plantation crops and produce such as cotton, tobacco, sugar and rum were sent to Europe, which in turn sent manufactured goods of various kinds to West Africa in exchange for slaves. But the Sahara still remained a formidable barrier, and the outlook of the Guinea and Sudanic states of West Africa which before had been northwards across the Sahara towards North Africa and the Middle East became reversed southwards towards the sea which provided new and much easier links with Europe and America. Even today, despite the development of air transport, the Sahara still remains a serious barrier to transport and communications, and most of Black Africa's trading links with the rest of the world are by sea.

The slave trade ended at the beginning of the nineteenth century and was replaced by what has been described as a more 'just and equitable traffic' (Hancock 1942) in minerals and other raw materials, notably palm oil, required to feed the growing industries of Europe. From merely trading with the peoples of Africa through a number of scattered toe-holds along the coast, the European nations began to turn their attention to the acquisition of actual territorial possessions on the continent. Thus began the 'scramble for Africa' and the emergence of colonies ruled by Europe.

Here again the physical geography of Africa played a very important role in shaping the course of events. Unlike North America and the greater part of South America, where climatic conditions favoured extensive European settlement on a permanent basis, Africa's climatic conditions and the prevalence of certain dangerous diseases, especially malaria, militated against such settlement except in specially favoured areas where either the latitude or the altitude helped to reduce the high temperatures and humidities common in most parts of the continent. Accordingly, white settlement came to be associated mainly with the highland areas of East and Southern Africa (Fig. 6).

In 1652 the Dutch had established a small settlement around Table Bay near the extreme southern tip of the continent in what is now South Africa. Initially the settlement was intended only as a victualling station for ships travelling from Holland to the East Indies, but in time it became a permanent settlement which grew in the course of the next three centuries to become the present-day Republic of South Africa. In the middle of the nineteenth century, by which time the initial colony had come under British rule, there was further expansion inland beyond the Limpopo River under the stimulus of Cecil Rhodes, resulting in the creation of two additional British possessions, Southern and Northern Rhodesia, both of which attracted considerable numbers of white settlers. In the meantime the Portuguese were also establishing colonies in Angola and Mozambique on the Atlantic and Indian Ocean sides of the southern half of the continent. By the end of the first decade of the present century the British operating through the Suez Canal had

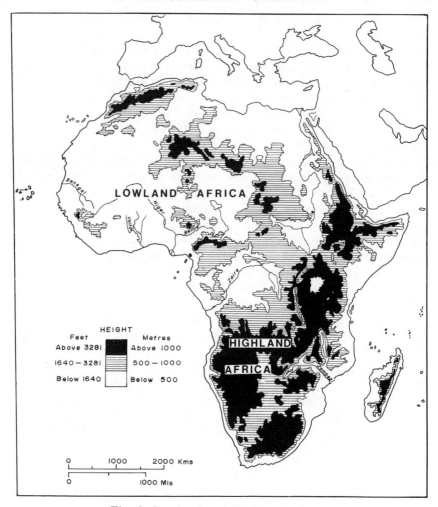

Fig. 6. Lowland and Highland Africa

established further colonies in Kenya and Nyasaland (now Malawi) for the settlement of Europeans, while the Germans had also established a colony in Tanganyika (now Tanzania) in addition to South West Africa immediately to the north of the Cape Province of South Africa.

This is not the place to go into the history of the European partition and colonisation of Africa. The point we wish to make here is simply that as a result of the orographical contrasts between the northern and southern halves of the continent, two distinct types of colonies came into existence in Africa: *colonies of settlement* in Highland Africa south of a line between the mouth of the Congo River and the Ethiopian massif, and *colonies of exploitation* to the north.

48

It was not altitude and climate alone that determined the types of colonies established in Africa; other important factors were the nature of the colonial policies followed by the individual European powers concerned and the prevailing social and political conditions in each particular area. Thus, although parts of the former Belgian Congo, such as Katanga and Ruanda–Urundi (now Rwanda and Burundi) were climatically suitable for white settlement, in fact no such settlement occurred because Belgian colonial policy did not favour the idea. Similarly, even though climatic conditions in Uganda were suitable for white settlement, as in Kenya, yet the British did not attempt any large-scale settlement there most probably because the country was much more effectively occupied and much more developed socially and politically than Kenya seemed to be at the time. On the other hand, in South Africa and Rhodesia, despite strong opposition from the local inhabitants, both the Boers and the British pushed through white settlement and actual territorial conquest, no doubt because of the strong attraction offered by the mineral wealth and agricultural potentialities of these areas. In North Africa, too, despite considerable local opposition from the Berber and Arab populations, the French acquired political control, notably in Algeria, and proceeded to establish large settlements for Frenchmen and other Europeans as a means of extending the political and economic spheres of France. Thus there was no clearly uniform pattern to white settlement throughout the continent. What can definitely be said is that in almost all cases wherever there seemed to be a chance and the prospects for economic or strategic gain were good, every effort was made to establish colonies of settlement rather than simply colonies of exploitation.

The partition of Africa effectively took place in the short space of time between the Berlin Conference in 1884–85 and the end of the nineteenth century. Up to the time of this conference most of the European nations with interests in Africa had been content with a few coastal footholds and vaguely defined hinterland spheres of influence. However, by enunciating the principle that in future no claim to territory in Africa by any European power would be recognised unless that power could show that it was in effective control of the territory concerned the Berlin Conference set in motion the scramble for Africa by the European powers. After the partition Africa became the largest colonial domain in the world with all except two of its countries, namely Ethiopia and Liberia, under foreign domination, and its peoples now freed from the threat of slavery were thrown into new cultural and political moulds imposed by Europe.

At the beginning of the 1914–18 war the main colonial powers in Africa were Britain, France, Belgium, Germany, Portugal, Spain and Italy. At the end of the war Germany was pushed out of the African scene by the victorious allies who shared or took over her colonies as Mandated Territories under the general auspices of the League of Nations. Thus, the colonial character of Africa continued unchanged, and at the outbreak of the Second World War in 1939 the only independent African states were Egypt and Liberia, Italy having annexed Ethiopia, Africa's oldest independent state, only a few years earlier by military conquest and turned it into a colony. It was not until after the war that the process of African emancipation began to gain momentum, first in North Africa and North-East Africa, and subsequently, starting with Ghana in 1957, in Black Africa south of the Sahara. At the beginning of 1978, the only parts of the continent that were still under alien rule were South Africa, Namibia, Rhodesia, the former Spanish Sahara now partitioned by Morocco and Mauritania, and a few Portuguese and Spanish enclaves and islands in North Africa.

Thanks to modern developments in transport and communications, the former isolation of Africa south of the Sahara has now been broken and practically all parts of the continent, especially the areas that border the sea, are today in effective contact with the rest of the world. Politically, also, the newly independent states of Africa are able to participate fully in world affairs through the United Nations, and their large numerical strength, which is a direct legacy from the European partition of the continent, has given these countries considerable voting power on world bodies far in excess of their real political and economic strength. However, it still remains a fact that in terms of sailing distance the southern and eastern parts of the continent are still somewhat remote from the world's principal economic foci, Europe and North America. While the Suez Canal is in operation things are not so bad for East Africa, but when it is closed all sea traffic from Europe and America must reach it via the much longer route round the Cape of Good Hope. Nevertheless, the availability of air transport has gone far to ameliorate the position.

As regards internal communications, also, much progress has been made in Africa during the past fifty years, although a great deal still remains to be done. In the humid tropics, animal transport has never been of much importance in Africa owing to the rifeness of the deadly tsetse fly. Horses, donkeys, and mules are quite common in North Africa and in the drier Sudan lands south of the Sahara, while in South Africa oxen played an important role during the nineteenth century when the Boers were pushing their frontier of settlement inland from the coast and the ox wagon was their principal means of transport. But by far the oldest known example of the use of animal transport in the continent is the camel, which was introduced as a beast of burden in the Sahara by the Arabs in the seventh century AD. This animal brought about a major revolution by enabling the Mediterranean coast of North

Africa and the Middle East generally to become linked by means of caravan routes with the Sudanic states along the southern borders of the Sahara. Regular camel caravans thus came to be established across the desert which have persisted to this day, though on a greatly diminished scale, and camels are also used widely in the countries of North Africa.

In specially favoured areas within the humid parts of the continent where animal transport was possible and available it played a vital role in the period before the advent of railways and motor transport. This point is well underlined by Mary Kingsley in her account of West Africa where she states that the British were forced to move their capital from Cape Coast to Accra in 1876 because, among other reasons, horses could thrive in the drier climate of Accra, which was not possible at Cape Coast because of its greater humidity and the dangers of tsetse infestation (Kingsley 1965). Even in Accra she observed that horses were safe only within a distance of twelve miles or so of the town.

For all practical purposes, then, the principal means of transport within the greater part of Africa, especially the tropical parts, before the advent of railways and the motor car were human transport and water transport. But each of these had serious limitations. Human transport was not very effective because of its slowness and the very limited amounts that could be carried per head. Only when slave labour was used could human transport be said to be effective; but even so the cost in misery and suffering far outweighed any advantages, and in the long term this form of transport was highly wasteful and expensive (Hailey 1956: Ch. xxiii).

Water transport would have been ideal for internal communication if long stretches of navigable waterways had been present; but as has already been stated this was not the case in many parts of the continent. Undoubtedly, the rivers and lakes throughout the continent were used for transportation, but except for the basin of the Congo River and along sections of the Nile, the Niger and its tributaries, the Gambia and on some of the Great Lakes of East Africa, the scale, as far as the rivers were concerned, was very limited and highly localised because of the abundance of rapids and other interruptions. On the sea transport was made difficult first of all by the paucity of sheltered bays and estuaries which could provide safe anchorages and also by the particularly dangerous nature of the surf along most of the coast. Thus, point-to-point sailing such as was commonly employed elsewhere in ancient times for coastwise traffic was largely ruled out, while the idea of long-distance voyages across the open seas was entirely out of the question owing to the absence of suitable craft for the purpose, the many physical dangers involved, and inadequate knowledge of navigation as well as of the winds and currents of the open seas.

The result of these difficulties was that, while Africa's links with the outside world were undergoing rapid improvement, within the continent itself most of the people lived in isolated communities often on uneasy terms with their neighbours. In Africa south of the Sahara, only in the savanna and steppe lands of the Sudan of West Africa where the open character of the vegetation makes transport relatively easy and a few other places in the continent do impressive empires embracing large groups of people appear to have developed. Consequently, when the European colonial powers started drawing boundaries across the continent to indicate the extent of their newly acquired possessions many different African peoples were forcibly brought together for the first time within common political units and obliged to develop new loyalties and new human associations. The forms of transport which particularly forwarded these new political developments and have become most notably associated with them are the railway and the motor car. They are among the most potent instruments of political change and economic development that have been in operation in Africa since the turn of the century.

Railways made their first appearance in Africa towards the end of the nineteenth century, about the same time that the partition of the continent was under way, while motor transport appeared some twenty years later, about the beginning of the First World War. Railways were used for two purposes in the early stages: firstly as instruments for military conquest, and secondly as instruments for economic exploitation. In view of the need at the time for the colonial powers to stake their claims to territory as rapidly as possible, many of the railways built then were hurriedly constructed on narrow gauges from coastal termini to inland points which had special strategic or economic importance. Minerals in particular played a very important role in promoting railway construction in West, Central, and Southern Africa, especially in the British territories. It was mostly in the French colonies of West Africa that strategic motives were particularly prominent.

The circumstances in which railways came to be built in Africa inevitably made for a series of segmented and uncoordinated systems all over the continent. Not only did many railways stop short at international boundaries, they were also built on a variety of gauges ranging from the very narrow 2 ft 6 in. gauge found most notably in Sierra Leone to the wider 4 ft 8 in. gauge found mainly in North Africa. It has been suggested that one reason for this multiplicity of gauges was the reluctance of the colonial powers, out of self-interest, to encourage the development of a uniform rail system that might lead to easy physical links with other territories in the continent. If this is so, they have been eminently successful; for even today, despite the fact that most of the countries of Africa are in control of their own affairs, the continent is as much segmented from the point of view of rail transport as it was during the colonial period.

If railways laid the foundations for the modern political and economic development of Africa, it was roads that made possible the realisation of tangible results at the grassroots level. Because of its great flexibility and its relative cheapness road transport has become the most widespread and important form of transport throughout the continent. But even here it cannot be said that a trans-continental system as yet exists, although in recent years various schemes for developing such a system have been discussed and a number of major improvements to certain key international road links have been undertaken. Even within individual African countries a great deal more needs to be done before the existing road systems can really be said to be adequate. Ideally, for every 26 sq. km (10 sq. miles) of land in settled or productive areas a minimum of 9.7 km (6 miles) of roads or railways, but preferably roads, is needed to provide a really effective system of transport and communications.

It may be helpful at this point to quote some figures showing the relative cost of human transport, rail transport and road transport. According to Lord Hailey (1956: p. 1,536), during the First World War the transportation of 4,200 tons of foodstuffs by the French was recorded as having involved the employment of no less than 125,000 carriers, while it was calculated in 1926 that in Nigeria the transport of one ton per mile by head porterage cost 2s 6d, by motor transport 1s, and by railway 2d. Again, according to the East Africa Royal Commission of 1933–35, whereas in East Africa the cost of transport by head porterage was 8s to 12s per ton mile, it was reduced by the use of motor transport to between 5s and 3s 6d.

Finally, we come to air transport, which is the latest form of transport to make its appearance in the continent. The main drawbacks of air transport are its very high cost and the highly technical infrastructure which its operation demands. As far as Africa is concerned, however, the chief value of this mode of transport lies in the ready links which it provides with other parts of the world and its ability to establish quick contacts within Africa itself between places that are unable to communicate easily or quickly with each other because of physical obstacles which make the use of other modes of transport unduly difficult or even impossible. Certainly, from a purely political point of view, air transport has brought the countries of Africa much closer together than at any previous time in history and given a new meaning to the concept of African Unity.

4

Colonialism in Africa

It would be impossible in a single chapter to deal adequately with the subject of colonialism in Africa even if we omitted consideration of its earlier phases and confined ourselves to its more recent manifestations. In this chapter, therefore, we shall deal only with modern European colonisation in the continent, although it must be remembered that other racial and ethnic groups, such as the Arabs and a number of African peoples have in the recent past engaged in the conquest and political control of other groups within the continent involving varying degrees of economic and other forms of exploitation. The Arabs in particular engaged extensively in slavery and wrought considerable havoc in certain parts of the continent, especially in East Africa, before and during the period of the European slave trade and actually established their rule over the islands of Zanzibar and Pemba and later along the whole of the East African coast from Mogadishu as far south as Cape Delgado from about the end of the seventeenth century until towards the end of the nineteenth century when their possessions came under European control. Without any doubt, all or most of these activities fall within the scope of the term 'colonialism', although the extent and intensity of their impact was much less than that of the modern colonising powers of Europe who dominated the African political scene in the nineteenth and twentieth centuries.

As we saw in the preceding chapter, all except four of the countries of Africa are today fully independent politically. With the sole exception of Liberia, which was colonised initially by freed slaves of African origin, all these countries have at one time or another been under European rule. Although European contacts in the greater part of the continent are of comparatively recent origin and actual political control by European nations, as distinct from purely trading activities, began in most cases only during the last few decades of the nineteenth century, following the Berlin Conference of 1884–85, yet European colonisation has left a strong imprint on the political and social life of the continent which no amount of cultural self-consciousness and nationalist sentiment in the newly independent states can obliterate completely. The legacy of the colonial past can be seen in practically every aspect of life;

but perhaps its most obvious and striking imprint is in the sphere of language. With the exception of North Africa, where Arabic is now the recognised official language everywhere, Ethiopia, where Amharic is now the dominant language, as it has been for centuries despite the growing importance of English, and Tanzania, where Swahili (no longer English) is now the official language, the official languages used in all the other countries of the continent are those of their former European rulers. Only in the case of South Africa, which in any case is not strictly an African country, has a completely new language, Afrikaans, which is peculiar to the dominant section of the ruling white minority, emerged as the official national language with English playing a subordinate role; although, even here it can be argued that Afrikaans has evolved from Dutch, which was the language of the original white settlers who founded the Cape Colony in 1652.

Because the European colonisation of Africa was by sea all the powers which established themselves in the continent were those with strong maritime traditions, led initially by Portugal and Spain, and subsequently joined by England, Holland, France, Denmark, Sweden and Germany. We have already seen how these nations confined their initial colonising efforts to trade in gold and other natural products like pepper, spices and ivory for which there was a great demand in Europe and how their attention shifted to trade in slaves after the establishment of plantations based on the use of cheap labour in the New World.

In no part of the continent was the struggle for territory and trading rights among the European nations more intense than in West Africa. With the exception of Italy, whose colonising efforts were practically confined to North and North East Africa, every one of the colonising nations had possessions in the area extending from between the Senegal River and the Congo and bordered on the north by the Sahara. Consequently, when the partition of the continent took place West Africa emerged as the most highly segmented area, even though by then a number of the smaller powers such as Denmark, Holland and Sweden had either withdrawn voluntarily or been forced out by their more powerful rivals. Within West Africa the area which attracted most attention from Europe and therefore became the scene of the greatest rivalry was the stretch of coast known as the Gold Coast, where in 1482 the Portuguese had built their first fort at Elmina (Fig. 7). Several factors were responsible for this. In the first place this part of Africa lay relatively close to Europe and America. Secondly, despite the inhospitable climate, it had an abundance of minerals and other natural resources, especially gold, ivory and palm oil. Thirdly, because of its relatively dense and generally virile population it served as a rich source of slaves who could be shipped conveniently to America. Finally, the direction and seasonal disposition of the prevailing winds as well as

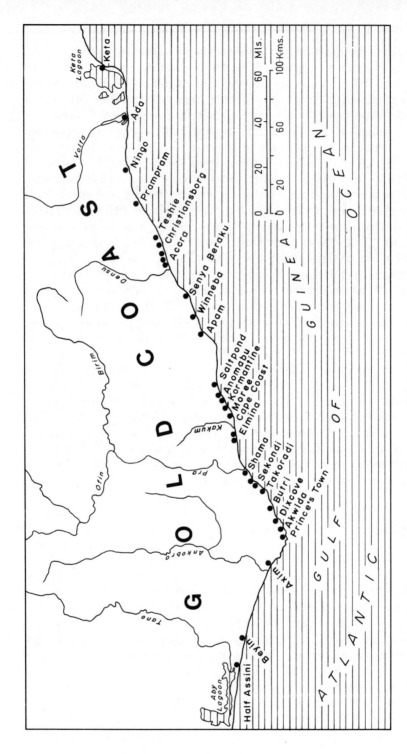

Fig. 7. Distribution of European forts along the Gold Coast between the fifteenth and nineteenth centuries

the ocean currents within the region and its adjoining seas facilitate the movement of the sailing ships which were then in use on the outward journey from Europe as well as on the return trip and also between West Africa and America, thus providing a convenient triangular arrangement for trading purposes in the North Atlantic which no other part of the African continent could match.

On the first part of the journey from Europe the ships moved from southern Portugal as far as Cape Verde with the aid of the Canaries Current and the North East Trades. In the region of Sierra Leone they encountered the Guinea Current which took them as far as the Gold Coast and beyond. Thus, provided they chose the right time they could make the journey in less than two months. Fortunately, the time when the North East Trades were strong enough to assist movement to West Africa was also the time when the climate, then dominated by the dry Harmattan, was at its best from the European point of view. The West African trading season was accordingly timed to begin between September and January and to end before the onset of the monsoonal rains in May. The return journey to Europe was rather more difficult, but, again, it was assisted by another ocean current, the Equatorial Counter Current, which also facilitated the movement of ships travelling across the Atlantic to Central and North America (Blake 1937).[1]

Despite all the advantages which West Africa had for trade with Europe and America there were many factors which militated against the actual establishment of colonies and plantations by Europeans within the area itself as had happened in tropical America. The first obstacle was the difficulties posed by the hot, humid climate of the coastal areas and the rifeness of dangerous tropical diseases, especially malaria, which combined to give the region its reputation as 'the white man's grave'. Next and no less serious was the hostility and opposition of the local population whose density and virility made them a much more formidable force than the weaker Indian populations of tropical America. The following moving report of how the King of Elmina in the Gold Coast greeted the Portuguese Captain, Diego de Azambuja, when he sought permission to build a fort on his land in 1481 is particularly instructive in this respect, even though the Portuguese eventually got their way:

considering the nature of so important a man as the captain and also of the gallant people who accompanied him, he perceived that men of such quality must always require things on a lavish scale; and, because the spirit of such a noble people would scarcely endure the poverty and simplicity of that savage land of Guinea, quarrels and passions might arise between them all; he asked him, therefore, to be pleased to depart, and to allow the ships

1 Although the combined effect of prevailing winds and ocean currents is theoretically plausible it is unlikely that the role of the ocean currents was quite as important as Blake suggests.

to come in the future as they had in the past, so that there would always be peace and concord between them. Friends who met occasionally remained better friends than if they were neighbours, on account of the nature of the human heart [Barros 1552].

For many centuries the European traders had to confine themselves to the establishment of coastal forts from which their trade with the coastal and inland peoples was conducted. Even so most of the forts were built on land that was leased rather than conquered and to a large extent the activities of the European traders were carried out on sufferance from the local population rather than by right (Oliver and Fage 1962). Attempts were made to establish plantations during the fifteenth and sixteenth centuries, but these proved unsuccessful with the result that it was only on the more accessible and manageable offshore islands, such as the Cape Verde Islands, and the islands of the Gulf of Guinea, particularly São Tomé, that the Portuguese succeeded in establishing such plantations. Thus, right up until the late nineteenth century the only really effective domain of the European trading nations in West Africa remained the high seas and the forts they controlled along the coast. Beyond these forts the only places on the land where they had any real power were the immediate peripheries of the forts within range of their guns. Elsewhere control was fully in the hands of the local populace themselves. The only place where inland penetration was seriously attempted during this earlier period was in the Senegal–Gambia region, where the French, taking advantage of the greater accessibility of the Sahel zone along the northern borders of West Africa, travelled some three hundred miles up the Senegal River, but without achieving much in the way of economic gain owing to the relative poverty of the area in both population and trade.

The dominant activity of the European nations in West Africa from the sixteenth to the nineteenth century was the slave trade, which was begun by the Portuguese in about 1510 and taken over almost exclusively by the Dutch after 1642 when they ousted the Portuguese from most of their coastal possessions. It has been estimated that by 1600 some 900,000 slaves from West Africa had been landed in the Americas, and that the numbers landed in the seventeenth, eighteenth and nineteenth centuries were, respectively, 2,750,000, 7,000,000 and 4,000,000 (Oliver and Fage 1962). The area which came especially to be associated with this nefarious trade was the stretch of coast between the Gold Coast and the Niger delta, which soon became known as the Slave Coast.

The slave trade was by no means confined to West Africa. In due course it spread to other parts of the continent, especially Angola and East Africa, where it caused even more serious devastation owing to the very fragile nature of the economies of these areas and the general sparsity of their populations. But even in West Africa, where a number of factors, such as the fecundity and natural resilience of the population, helped to diminish its effects very considerably, the trade caused extensive and irreparable damage to the entire fabric of society.

The political penetration and conquest of Africa by the European powers was assisted considerably before and during the process of partition by the work of explorers bent on unravelling the geographical mysteries of the continent's unknown interior as well as by the work of missionaries who were convinced that the surest means of banishing slavery and of bringing peace and stability to the continent was by the conversion of the inhabitants to Christianity. In many instances, although the basic motivations were different, political and missionary activities proceeded hand in hand and complemented one another. It is thus an over-simplification to suggest that the sole motive for the establishment of colonies was in all cases pure economic exploitation aimed at benefiting only the power concerned. Undoubtedly, economic gain was almost invariably the dominant motive, but philanthropy and moral upliftment also constituted important strands in the policies of the colonising nations (Busia 1962). In particular, especially in Britain, increasing concern was felt about the evils of the slave trade and the need for its abolition despite the opposition of many of the European nations and the hardened African middlemen who profited from it.

Attempts to abolish the trade began in Britain in the middle of the eighteenth century, but it was not until 1807 that the British parliament actually passed a law making the trade illegal for British subjects and some fifty years later that it was finally outlawed in all the civilised parts of the world. Following this important development, the whole basis of European trading activities and of the European presence in West Africa took on a new character, and tropical Africa, especially West Africa, once again resumed its former role as a source of raw materials which were now required in even greater quantities for the rapidly expanding industries of Western Europe. Unlike slavery, the successful pursuit of this new trade called for both the investment of capital and the maintenance of effective political control by the European powers who still remained in the continent. These were the considerations that eventually led to the partition of the continent. By the end of the nineteenth century Africa was without any question the most important as well as the most extensive colonial domain in the whole world.

At this time the European powers with possessions in Africa were Britain, France, Germany, Portugal, Spain, Belgium and Italy, and the disposition of their territories was as follows: In North Africa the French controlled Algeria, Morocco and Tunisia; Britain held Egypt; Italy held Libya, while Spain held the small possessions of Ceuta, Melilla and Ifni. In West Africa Britain held The Gambia, Sierra

Leone, the Gold Coast and Nigeria; France held Mauritania, Senegal, Guinea, French Sudan (now Mali), Niger, Ivory Coast, Upper Volta and Dahomey; Portugal held Guinea and the Cape Verde Islands; Spain held the Sahara Province, while Germany held Togoland. In Central Africa Germany held the Cameroons and Ruanda–Urundi; Belgium held the Congo; France held Equatorial Africa and Chad; Portugal held São Tomé and Principe, while Spain held Rio Muni, Fernando Po and Annobon. In North East Africa Britain held the Sudan and part of Somaliland; Italy held another part of Somaliland as well as Eritrea, while the French held yet another part of Somaliland. In East Africa Britain held Uganda and Kenya, including the islands of Zanzibar and Pemba, while Germany held Tanganyika. In Southern Africa Britain held South Africa, the three protectorates of Basutoland, Swaziland and Bechuanaland, Northern and Southern Rhodesia, and Nyasaland; Portugal held Angola and Mozambique; Germany held South West Africa and France held the island of Madagascar.

The largest colonial powers in the continent after 1900 were Britain, France, Belgium and Portugal. Germany was also important, but her empire was very short-lived; for, as has already been pointed out, she lost all her possessions to the victorious allied powers after her defeat at the end of the First World War in 1918. France's influence was dominant in North, West and Central Africa, Britain's in West, Southern and East Africa, Portugal's in Southern Africa, while Belgium's influence was confined to her large possession in Central Africa.

In the political control and administration of their respective African empires the various European powers followed different colonial policies depending, as has been suggested earlier, on the prevailing local circumstances and the particular attitudes and political philosophies of the powers concerned. Sometimes, as in the case of the British, it was difficult to discern any clearly defined policies except in a very general sense; things were left to take their natural course and policies were made in accordance with the dictates of changing situations at home and abroad.

COLONIAL POLICIES OF THE EUROPEAN POWERS

Britain

British colonial policy in Africa, as has already been indicated, was not consistent throughout the continent. It took different forms in the lowland parts of Africa where there was no permanent white settlement and in the highland parts where settled white minorities were established at quite an early date. In the white areas the assumption was that new British outposts would develop more or less along the lines of the white British dominions in Canada, Australia and New Zealand. The only serious complication in Africa was the presence of large numbers of Africans who in all cases formed a majority of the population. Accordingly, the aim of British policy was to find means of safeguarding the interests of the African populations and those of non-white immigrants, especially Indians, within a system of government based on the political and economic dominance of the white minorities. Accordingly, as these countries gained in political maturity and advanced towards self-rule Britain sought by means of various entrenched clauses in their constitutions to safeguard the fundamental rights and basic interests of the African populations, but without overtly questioning the right of the white minorities to assume eventual political dominance.

South Africa was the first of the white colonies to gain self-government, and it soon became clear that the safeguards entrenched in the constitution were of no avail in protecting the political and economic interests of the non-European population. In spite of this Britain continued to hold optimistic views about the goodwill and fair-mindedness of white minorities in other colonies in Southern and East Africa, notably Rhodesia and Kenya, and no firm measures were taken to prevent the repetition of future breaches of faith similar to what had occurred in South Africa. British policy thus continued to be based on vague and optimistic notions about the possibility of racial partnership and cooperation in these countries, despite all the ominous pointers to the contrary. Even here partnership and equality were viewed as long-term goals to be achieved gradually as the non-white populations moved upward on the educational and economic ladder. While the idea of complete separation between the races was officially abhorred, yet in practice the manner in which land and political and economic power were apportioned among the races inevitably seemed at best to point to the eventual development of plural rather than fully integrated societies in these countries.

In the lowland parts of Africa where the complications of white settlement did not exist British colonial policy was more straightforward and definite. Policy here was summarised by the phrase 'indirect rule', an idea first put forward in its clearest terms by Lord Lugard in 1926 in his *Dual Mandate in British Tropical Africa* and implemented in its most successful form by Sir Donald Cameron during his governorship of Tanganyika. Indirect rule was based on the theory that the colonial possessions were territories held in trust and that consequently it was Britain's obligation to help the colonial people to advance and to develop the resources of the colonies not for the exclusive benefit of Britain but of mankind generally. In administering these possessions, therefore, Britain sought to interfere as little as possible with the traditional native institutions but rather to improve and use them for the implementation of governmental policies. It was assumed that ultimately the African colonies would stand on their own feet, although no clear conception seemed to exist as to the precise constitutional forms which the process towards eventual autonomy would follow or when the final goal would be reached. What was clearly not envisaged was that there should be eventual political assimilation with Britain or that the inhabitants of these colonies should be transformed into black Englishmen detached from their indigenous African roots. In pursuance of this policy considerable attention was paid to the provision of education and of health and social services although it was left to each individual colony to determine the pace of its own development, depending on its economic resources and the willingness and initiative of its inhabitants and of the particular administrators in charge of affairs.

It must be pointed out, however, that in the framing of these policies Britain was not an entirely free agent. In a number of the African colonies themselves, especially the Gold Coast, there was strong local opposition to foreign domination and the imposition of foreign institutions, and the British administrators, even when their intentions were unimpeachable, had to tread warily in case they outraged local opinion led by the chiefs and the educated intelligentsia.

By the end of the Second World War, however, British colonial policy in Africa had moved one stage further. It was now fully accepted that Britain had the responsibility of 'promoting economic and social advance in her colonies in order to provide the essential basis for political self-rule' (Hailey 1956: Ch. v). This new policy began to be vigorously pursued in West Africa, but in East and Southern Africa outside of the Union of South Africa it also began to lead to serious official misgivings about the policy of permanent white settlement in the countries which were still groping their way towards a satisfactory formula for self-government based on racial justice.

The justification for the policy of indirect rule, especially in West Africa where it had been practised most faithfully and successfully, came when the time for self-government arrived. The transition proved remarkably smooth and Africans were able to take over from the British administrators in practically every sphere of life. The only things that proved to be outstandingly lacking within the respective colonies were a sense of genuine national unity and a properly and widely established tradition of democratic government.

France

In contrast with the British approach, French colonial policy was more definite and pragmatic in its aims and methods. It was based on two principles – direct rule and political assimilation. The French believed themselves to be the heirs of the Roman tradition of empire and saw their mission as that of a superior race with a duty to extend the benefits of their civilisation to the backward inhabitants of their colonies and to reward them with French citizenship when they showed sufficient evidence of having embraced this civilization.

In pursuance of these policies little attention was paid to indigenous native institutions, which the French generally despised, and a system of government based on French ideas was imposed everywhere. All authority was centralised in Paris, from where it radiated through the local governors and assemblies down to the level of the people. Two particularly notable features of the system were the imposition of a poll tax as well as forced labour on all the adult population who did not have the highly coveted status of full French citizenship.

The idea of assimilation was based on the revolutionary doctrine of the equality of man and the assumption that French culture was superior to that of the African, while the policy of direct rule was a necessary adjunct of the policy of assimilation. It appears, however, that the French never really faced the full implications of assimilation and in practice very little was done to give effect to it (Crowder 1962). In the early zeal of colonisation full citizenship and the franchise were given wholesale to the Senegalese of the coastal towns as far back as 1848, but with the expansion of the empire and the inclusion within it of much less-developed areas and peoples this policy was replaced by one of association in which only a few select individuals were accorded citizenship. Thus in 1921, twenty years after the establishment of the French West African empire, the vast mass of the African population had the inferior status of 'sujet' (subject people), which made them liable to summary administrative justice and forced labour, while only about 555 people, outside of Senegal, enjoyed citizenship. In 1939, out of a total population of 15 million in the French West African empire there were only 2,136 citizens outside of the four Senegalese towns of Dakar, Goree, St Louis and Rufisque (Crowder 1962).

In the early days of their West African empire the French had sought to involve the local chiefs in the administration, but were quickly disillusioned by the internal dissensions among these chiefs and their general incapacity and cruelty in their dealings with their subjects. Consequently, only a few whose loyalty could be counted upon were eventually retained, but even these were reduced to the status of mere functionaries whose main role was the collection of taxes, the recruitment of labourers and soldiers, and the settling of minor disputes. Thus, unwittingly, the French undermined an institution which had traditionally been an undivided religious, economic and political source of authority and replaced it with a foreign one without any local roots.

Wartime exigencies, however, brought about important changes in policy. The Brazzaville Conference of 1944 proposed a federation of the colonies with France, but the Constituent Assembly of 1945 held immediately after the war proposed a Union instead. Forced labour and the status of 'sujet' as well as a number of other restrictive laws were abolished and the colonies were later given increased control over their own affairs. The proposed Union was to consist of Metropolitan France and Overseas Departments and Territories functioning under the presidency of the President of France. In 1946 French citizenship was extended to all French colonial subjects, but in practice the franchise was limited to those whose political education was considered to be sufficient.

While many of the African elite were impressed by these changes and took pride in the close political and cultural identity which they shared with France, the rising tide of nationalist consciousness began to create growing disenchantment with the whole idea of assimilation and association within a political Union controlled by France. Yet because of the thoroughness with which France had pursued her policy of direct rule it was found that a great deal of damage had been done to most of the indigenous political and social institutions required for the development of authentic African nations. Another serious defect which became apparent was that in their development programmes the French had concentrated most of their attention in the urban centres where French administrators and the African elite were to be found, while the rural areas, which constituted the bulk of the empire, had been grossly neglected. And because the idea of eventual self-government for the African colonies had never entered into the thinking of the French, when independence eventually came hardly any of the colonies, with the possible exception of Senegal, possessed the trained manpower and infrastructural support needed to guarantee a smooth and easy transition, and the change was therefore a painful wrench.

Like the British, the French also had colonies of settlement in Africa. These were the North African colonies of Algeria, Morocco and Tunisia. In Morocco and Tunisia the numbers of French settlers were not so large; but in Algeria, which for nearly a century had been administered as a virtual part of France, there were in 1962 about 400,000 French settlers who were in possession of vast tracts of agricultural land. Here, the French had no illusions whatsoever about the possibility of eventual self-government for the local populace and no attempt was made to devise political institutions for such an eventuality. When at last the Arab population began to agitate seriously for independence, France's attempts at the granting of concessions were vehemently resisted by the white settlers who had come to regard the country as their rightful and exclusive possession, and it was only after a long and bitter war that the Algerians succeeded in wresting their freedom from France.

Portugal

Portugal was the first colonial power in Africa and the guiding principles of her colonial policy in Africa were assimilation and paternalism. The Portuguese were ·noted for their ruthlessness in their dealings with the local inhabitants. Although dispossessed of most of their coastal strong-holds in Africa, they were able to retain a number of possessions which were later extended to form the subsequent Portuguese empire.

Because of her decline as a world power and her relative poverty among the countries of Europe, Portugal was not able to accomplish much in her colonies. Far from helping her colonies to stand on their own feet, Portugal sought to exploit them for the benefit of her own fragile economy. In consequence the colonial· possessions were regarded as overseas extensions of the metropolitan country and the local inhabitants were given very little say in the conduct of affairs. In pursuance of the policy of assimilation token attempts were made to accord Portuguese citizenship to a handful of the colonial peoples, usually those with a mixture of Portuguese blood, while the rest of the population were accorded very few political rights and privileges. The policy of paternalism expressed itself in Portugal's belief that it was her sacred duty to call forth to civilisation her African subjects from their present state of human degradation by freely exploiting their labour for the benefit of the mother country until such time as they were deemed fit to take their place alongside the Portuguese as full citizens of the empire. Right from the outset the importance of Christianity was stressed as a means of providing spiritual and moral upliftment for the colonial subjects, but the humanitarian aspects of the Christian religion appear to have been completely ignored by the Portuguese government in its actual dealings with the colonial peoples.

It would appear then that both the policy of assimilation and paternal responsibility were nothing more than myths used to conceal Portugal's cruel and naked exploitation of her helpless colonial subjects. Despite the boast about Portugal's non-racial, Christian civilising mission in her colonies, official Portuguese sources disclosed in 1950 that there were 1,478 civilised and 502,457 uncivilised Africans in Portuguese Guinea, 30,089 civilised and 4,006,598 uncivilised in Angola, and 25,149 civilised and 5,646,957 uncivilised in Mozambique, while it was reported in 1961 that of the $10\frac{1}{2}$ million people in Angola and Mozambique, Portugal's largest and richest colonies in Africa, over 99 per cent were illiterate (Duffy 1962).

The fact is that although the policy of assimilation was genuinely practised for a few years at the outset of the colonial empire, it was soon abandoned. Up to 1926 various half-hearted attempts were made to give the colonies, notably Mozambique and Angola, a measure of autonomy in the management of their domestic affairs, but these were abandoned and in 1930 they were declared to be part of the metropolitan country. In 1951, however, they were redesignated as overseas provinces of Portugal and were assumed, therefore, to have shed their colonial status. In reality, however, little change occurred in their status or their mode of administration. As a means of providing economic relief for the metropolitan country Portugal sought to settle large numbers of Portuguese in these overseas territories, especially Angola, and employed both force and dubious legal devices to secure cheap labour supplied by the African population for the development of vast European-owned estates and industrial concerns, while making pious pronouncements about the sacredness of her colonising mission and the benefits it was calculated to bring to the colonial peoples.

That the colonial subjects were not impressed by these pronouncements is clearly borne out by the fact that on 15 March 1961 rebellion against the Portuguese government suddenly erupted in the northern part of Angola and spread to Mozambique and Guinea until these countries gained their independence between 1974 and 1975.

Spain

Although Spain was a close rival of Portugal in the early days of European overseas colonisation, she never became a major colonial power in Africa. The reason is quite simple: in 1441 exclusive rights in West Africa, which was then the main focus of interest, were reserved to the Portuguese by a Papal Bull, and in 1494 the Treaty of Tordesillas further confirmed and extended these rights to the whole of Africa and the Asiatic mainland, while Spain was given similar rights in most of the new world, which had then just been discovered (Harrison Church 1956). Thus, in the Gold Coast, for example, where European colonising activity on the west coast of Africa was most intense during that period, Spain did not feature at all, although she was able later on to gain control of the islands of Fernando Po, Annobon, Corisco and the Elobeys situated in the Bight of Biafra farther east, and of the small mainland territory of Rio Muni. Elsewhere, apart from the Canary Islands off the north-western coast of the continent which she acquired in 1497, her only colonial possessions in Africa were the Saharan territory of Rio de Oro, now known more commonly as Spanish Sahara (now Western Sahara), together with the small enclaves of Ifni, Melilla and Ceuta, farther north.

Because of the comparatively small size of her possessions and their relatively sparse populations little was heard about Spanish colonial policy. In many respects the features of this policy were very similar to those of Portugal, that is, assimilation and paternalism, although there was much less pretence about the benefits of Spanish rule for the subject peoples. Politically and economically, all these possessions were treated either as provinces or as appendages of Spain and ruled more or less directly from Europe. But Spain was even more ruthless than Portugal in suppressing local opposition and in directing the resources of her colonies towards the mother country. Nevertheless, in October 1968 Spain granted independence to Equatorial Guinea comprising Rio Muni and the neighbouring islands over which she had for many years had a rather tenuous hold; but remained firmly in control of Spanish Sahara until early in 1976 when she transferred it to Morocco and Mauritania. The Canary Islands, however, are still administered as an overseas province of Spain.

Like the Portuguese, the Spanish in their treatment of their African subjects made a distinction between those who were culturally 'evolved' and those who were not. The evolved Africans were divided into two classes – those who were fully evolved or the *emancipados* and those who were only partially evolved or had had the benefit only of *emancipacion limitada*. The former enjoyed the same civil rights as Europeans and were subjected to the metropolitan penal code, while the latter had only restricted civil rights and came under the jurisdiction of native tribunals. Lastly, came the mass of the population who were not emancipated and who enjoyed few civil privileges and were wholly subjected to the jurisdiction of native tribunals (Hailey 1956: Ch. v). As far as the exercise of political rights was concerned, however, all the colonial subjects were equally denied any effective share.

Belgium

Belgium was a comparatively late arrival on the African colonial scene, her formal connection with the continent dating only from 1908, when she was compelled to annex the territory of the Congo Free State, which had previously belonged virtually as a personal estate to the King of Belgium, Leopold II, whose rule in the Congo had shocked and scandalised the civilised world on account of the shameless exploitation and the barbarous acts of cruelty against the native population with which it was associated. Being a small country located in a very sensitive part of Europe, Belgium made a conscious effort to administer her vast possession in such a way as to cause the least offence to the world powers while at the same time promoting her own economic interests.

The policy which Belgium adopted in the discharge of her colonial responsibilities was one of paternalism and a high degree of administrative centralisation based on Brussels. While she recognised that her colonial subjects would one day qualify for self-government, she did not consider it her duty to hurry that day but rather to concentrate in the meantime on economic and social development aimed at improving the material well-being of the colonial subjects. Because of the immense size of the Congo, the diversity of its peoples and the great abundance of its natural resources, especially minerals, the need was felt for a tightly organised administrative structure capable of exercising direct control in all spheres of activity. While attempts were made to involve the local populace through their chiefs in local government, little was done to give them a share in the central government.

In the field of education a great deal was done, often through the Christian Missions, but the emphasis was placed on elementary and technical education, while secondary and higher education were grossly neglected, the aim being to train artisans and other semi-skilled workers who could provide the manpower support needed for the exploitation and development of the country's vast economic resources. Thus, until only a few years before independence all the higher and even the middle positions in the civil service and private business remained a virtual monopoly of Belgians.

Like the French, the Belgians also accorded certain privileges to Africans who were considered to be 'evolved' on educational and cultural grounds, but these privileges did not include any effective participation in the government of the Congo, and it completely excluded any participation in the government of Belgium itself. In a sense the same limitations applied to the large numbers of Belgians living in the Congo either as civil servants or as private businessmen. Neither was large-scale permanent white settlement encouraged even in those parts of the country, such as the Katanga and Ruanda–Urundi, where climatic conditions were favourable. Socially, however, even though nothing was laid down officially, fairly rigid racial barriers were observed, especially in the large urban centres and in those areas with sizeable European populations.

After the Second World War a marked change took place in Belgium's attitude towards the Congo, and in the spirit of the United Nations Charter a number of important reforms began to be instituted. Between 1945 and 1947 social reforms, especially in the field of labour relations, were greatly stepped up. From 1947 to 1954 the main emphasis was on increased African participation in the government of the country, but the old paternalistic doctrines of assimilation based on the fulfilment by Africans of certain rigid educational and social qualifications reasserted themselves and little real progress was made. Then followed a period of gradual political emancipation from 1954 to 1958 in which determined efforts were made to democratise local government, to eliminate racial discrimination and to improve the quality of education, including the establishment of universities. But even these reforms proved inadequate to the growing expectations of nationalism, and within two years the initiative was taken out of the hands of Belgium and the country was plunged headlong into independence (Brausch 1961. See also Hailey 1956: Ch. v).

Throughout the period of Belgian rule in the Congo an interesting feature of official policy was the readiness with which business concerns belonging to or controlled by several of the major powers were allowed to operate in the Congo, in contrast with the more exclusive policies adopted by some of these same powers in their own colonial possessions in other parts of the continent. This was no doubt due to the fact that right from the creation of the original Congo Free State many of these powers had shown a strong desire to participate in the exploitation of the country's rich resources, but it was also due in part to a recognition by Belgium that the best means of guaranteeing her continued and unhampered control of this rich possession was to placate the larger powers who might otherwise cast envious eyes upon her vast African bonanza. Accordingly, in addition to pursuing an open-door policy in respect of the Congo she spared no efforts to keep the rest of the world informed about the country and Belgium's contribution to its economic development. Subsequent events were to show, however, that in spite of all that she had accomplished in terms of material development, there was still a great deal of discontentment among the Congolese with the way the country was governed and the marginal role assigned to them in the conduct of affairs.

Germany

Germany's entry into Africa began about the time of the Berlin Conference, but although she acquired large possessions in West, Southern and East Africa, she lost all of them to the Allied powers at the end of the First World War. These territories were Togoland and Cameroons in West Africa, South West Africa in Southern Africa and Tanganyika in East Africa.

Like the British, the Germans did not envisage the establishment of permanent white settlements in their West African empire, but they made attempts to establish such settlements in South West Africa and in Tanganyika, even though the actual scale of settlement was not particularly large. Everywhere, however, their colonial policy was marked by systematic economic exploitation and the ruthless suppression of all local opposition. The best example of this ruthlessness occurred in Tanganyika, where a local uprising known as the Maji Maji rebellion in 1905 sparked off by African resentment at the manner in which the Germans ignored local traditions, employed forced labour and compelled the African population to grow cotton was brutally crushed by methods reminiscent of the old Roman method of decimation. Earlier, in 1904, an uprising by the Herero tribe of South West Africa had been crushed with similar ruthlessness and the tribe reduced from 80,000 to 15,000.

Following these experiences German colonial policy began to undergo a certain measure of liberalisation; but the First World War brought Germany's colonial experiment to an abrupt end and her possessions were taken over by the victorious allies as Mandated Territories under the League of Nations. Because of the very brief span of her African empire, it is difficult to make a proper assessment of Germany's colonial policies. It appears, however, that despite their ruthlessness towards their subjects the Germans made determined efforts to establish a strong infrastructure for economic development in the form of railways and roads and by offering the local inhabitants sound and rigorous training in the basic vocations and crafts, the effects of which still persist in their former colonies.

Italy

Italy's colonial empire in Africa was much more restricted than that of Germany and almost as brief in duration. Up till 1935, when Italy forcibly acquired Ethiopia by waging a full-scale war against the country, the only parts of the African continent controlled by her were Eritrea, Italian Somaliland, both acquired about 1800, and Libya, which was annexed from Turkey in 1912 and which she made determined efforts to develop and colonise as an outlet for her surplus population.

It was largely the same motives that led to her conquest and colonisation of Ethiopia; but here her rule was even more brief than in Libya, which fell to the Allies in 1943 and was subsequently granted independence by the United Nations in 1951. Italy's rule in Ethiopia came to an end in 1941 with her defeat by the British in the Second World War.

A notable aspect of Italy's colonial policy in Africa was the brutal treatment which the Italian rulers meted out to their African subjects, especially where, as in the case of Libya and Ethiopia, the intention was to establish white settlements. However, precisely because of this intention and also because of the difficult physical environments in which they had to operate, the Italians spent considerable sums of money on the construction of roads and the development of ambitious irrigation works in their colonies, particularly in Libya, although they did very little for the social and political advancement of the local inhabitants.

CHANGING CONCEPTS OF COLONIALISM

We have already seen the changes which occurred in the attitudes of the European powers towards their African possessions between the fifteenth and the nineteenth centuries. Up to the beginning of the Second World War, however, the tide of colonialism was still riding high and no stigma attached to the possession of colonies, although the anti-slavery movement of the late eighteenth and early nineteenth centuries had brought about important changes in outlook, especially towards the colonial peoples.

Humanitarian sentiments similar to those that had inspired the anti-slavery movement asserted themselves again in the first decade of the present century when the barbarities being perpetrated in the personal empire of King Leopold II of Belgium in the Congo came to light and the ownership of the Congo was transferred to the people of Belgium under pressure from other Western countries. But the censure was not against colonialism as such but against the methods of a particular colonial authority. As for the philosophy of empire itself, the only significant change it had undergone up to the time of the First World War was to be found in Joseph Chamberlain's famous pronouncement that the colonies of Britain were henceforth to be regarded as 'imperial estates' to be developed by the mother country for her own benefit as well as that of the colonial subjects themselves and of the world in general.[2] Despite the many misinterpretations which Chamberlain's ideas have suffered (Hancock 1942), they clearly represented an advance on the view expressed earlier in the General Act of the Berlin Conference of 1884–85 that the principal obligation of the colonial powers was to bring the subject peoples the benefits of civilisation. Nevertheless, the general approach of most of the colonial powers in Africa up to this point was basically one of paternalism; the thought that colonial countries inhabited by non-white peoples should be prepared for eventual self-government hardly entered into the minds of the Western colonial powers.

The first important break from this frame of mind occurred at the end of the First World War, when mandates were set up. The relevant articles in the Covenant of the League of Nations stated that in regard to colonies and territories 'which have ceased to be under the sovereignty of the states which formerly governed them and which are inhabited by peoples not yet able to stand by themselves ... there should be applied the principle that the well-being and development of such peoples form a sacred trust of civilisation' and that the best means of giving effect to this principle is to entrust the tutelage of such peoples to advanced nations able to undertake this responsibility. Thus, for the first time, the government of certain colonial territories was made an international responsibility and the actions of the powers responsible for their affairs came under world scrutiny. And if such a development

2 *Parliamentary Debates*, House of Commons, 22 August 1895; and also *The Times*, 24 August 1895.

was possible in respect of certain of the colonial possessions, why not in respect of all the rest?

The Second World War took the idea of self-government for the colonies several steps further. During the war one of the main slogans of the Western Allies was that they were fighting for 'world freedom', and when at the end of the war the Charter of the United Nations was drawn up, the obligations of the colonial powers towards their subjects were underlined and 'equal rights' and 'self-determination' accepted for all peoples of the world regardless of colour or race or creed. In one of its most historic clauses the Charter declared:

Members of the United Nations which have or assume responsibilities for the administration of territories whose peoples have not yet attained a full measure of self-government recognise the principle that the interests of the inhabitants of these territories are paramount, and accept as a sacred trust the obligation to promote to the utmost, within the system of international peace and security established by the present Charter, the well-being of the inhabitants of these territories, and, to this end ... to develop self-government, to take due account of the political aspirations of the peoples, and to assist them in the progressive development of their free political institutions, according to the particular circumstances of each territory and its peoples and their varying stages of advancement [*United Nations Charter*, Article 73].

Many reasons can be assigned for this sudden upsurge of altruism. It is likely that one reason why the Western powers were anxious to speed up independence for their colonial subjects was in order to remove the whole question of colonies from the sphere of their own domestic politics and turn it into an international responsibility. The colonial question had become a bitter issue in the inter-war period, and Hitler had used Germany's lack of colonies as one of his chief grievances against the rest of the world. Japan had also done the same and had actually succeeded for a brief period as a result of her brilliant military campaign in the Far East in bringing many former European colonies under her rule. There was also the problem of India, once the biggest and brightest diamond in Britain's imperial crown but in the inter-war period and also during the war a veritable thorn in Britain's flesh. As the years went by the grounds for keeping India as a colonial possession became increasingly difficult to justify.

The post-war period also saw the exertion of new pressures on all the colonial powers by their subjects for self-government. This was due to several factors, including most notably the great improvements in communications which had rendered the former compartmentalisation of the world obsolete and brought all parts of the colonial world within much closer reach of each other; as well as the fact that many of the colonial subjects had fought in the war alongside troops from the metropolitan countries of the West, thus greatly broadening their outlook and strengthening their self-confidence and their sense of equality with their former colonial masters. In this changed situation few countries were in a position any longer to hold down unwilling subjects, and the two great powers in the post-war period, the United States and the Soviet Union, were not prepared, for different reasons, to lend the weight of their political and moral support to the perpetuation of colonialism in its traditional forms. There followed a rapid process of de-colonisation which in Black Africa began with Ghana's attainment of independence in 1957 and reached its culmination in 1960, when no less than seventeen African countries gained their independence (Boateng 1973).

THE COLONIAL BALANCE SHEET

During the inter-war years and particularly during the Second World War a considerable amount of introspection took place among the colonial powers regarding the value of colonies. It was argued in some quarters that colonies were serious liabilities for the controlling powers, while in others it was conceded that they offered both advantages and disadvantages (Ward 1973 and Walker 1945). It is obviously not easy to give a simple answer to the question whether colonies were assets or liabilities for those who owned them; but looking at the matter today and from the point of view of those who experienced colonialism as subject people, it is difficult to accept the verdict that colonial possessions were nothing but liabilities for the controlling powers, even if we credit these powers with the greatest degree of altruism in their quest for overseas possessions.

In attempting an objective assessment of the real position we need to make a distinction between the tropical colonies on the one hand and, on the other, the mid-latitude and temperate colonies, such as those in North and South America, Australia and New Zealand which were settled by Europeans right from the start and quickly developed into self-governing dominions and republics. The history of this latter group of colonies provides abundant evidence that they could not at any time have been serious liabilities to the countries which founded them originally, except perhaps in a political sense at the time when they were striving to achieve their independence. It is only in respect of the tropical colonies where permanent white settlement was not possible that the question whether they were assets or liabilities to the ruling powers is really valid. Even here the issues are not entirely clear-cut, for these colonies include those which were acquired for their strategic value, those which were acquired for their economic value, and those which were acquired for both these reasons.

In the case of purely strategic colonies there could be no question but that their ownership was advantageous, at least at the time of their acquisition. Had this not been the case they would not have been acquired in the first place and subsequently retained. Admittedly, strategic advantages are subject to change in the light of changing political and military developments in the rest of the world. But precisely because of this one would expect such colonies to be given up the moment they ceased to be advantageous to the controlling power.

It is in respect of colonies acquired wholly or primarily for economic reasons that one finds it really difficult to answer the question whether their ownership was an asset or a liability. Take the British colonies, for example. Most of these had begun as trading posts initially established by private companies. In India the commercial motive prevailed throughout, and the transition from Company rule to rule by the British Government was relatively smooth. In the case of the West African colonies, however, it happened that just when the most profitable trade under private control was beginning to end as a result of the abolition of the slave trade, the British government took over and was obliged to embark on a number of costly social ventures aimed at improving the lot of the inhabitants and at providing a reasonable infrastructure for more legitimate forms of trade. And these developments occurred at a time when the hostility of the local inhabitants was becoming more vocal and making life increasingly difficult for the British authorities. Economic advantages or not, it is on record that on a number of occasions during the nineteenth century the British parliament seriously considered the possibility of total withdrawal, and it was mainly because of humanitarian considerations and pleas from men, like Brodie Cruickshank, who were charged with local administration in the Gold Coast and who felt a strong concern for the welfare of the African peoples and the benefits which Christianity and Western civilisation could offer them, that Britain was finally persuaded to continue her rule in the country (Cruickshank 1853). The Gold Coast was by no means an isolated example; elsewhere the British were beginning to feel similar doubts about the future of their colonial rule in tropical Africa.

For the next fifty years the main effort of government was directed towards economic and social development in the African colonies, and missionary activity played an important supplementary role to the work of government officials. Considerable amounts of British capital flowed into the colonies for the purpose of promoting mineral exploitation and agricultural production aimed primarily at benefiting British investors; but actual government expenditure on administration and social services for the welfare of the colonial peoples was strictly related to the revenues which each colony was able to raise from its own resources. If the colonies had been abandoned at this juncture the ruling country would not have been seriously harmed; but the fact was that both sides, Britain as well as the colonies, stood to gain in different ways from the continuation of the association. As time went on, however, the investments made by the British in many of their colonies began to yield handsome dividends, especially in the case of those colonies with rich mineral resources. The colonies bore the cost of the expensive infrastructural services like railways and roads, while most of the profits from investments went to British investors at home.

In addition to these purely material gains, another important advantage which accrued to the colonial powers from their possessions was that of prestige. At a time when every European nation which was in a position to do so was acquiring colonies, it certainly seemed desirable to have as large and as rich a colonial empire as possible. Considerations of prestige seemed especially prominent among France's imperial motives. In 1870 France had suffered a humiliating military defeat at the hands of Germany, and it has been suggested that she sought to make up for this defeat by building up a strong overseas empire, especially in West Africa. Apart from the prestige of a strong overseas empire, France set about deliberately to develop this empire economically as a source of military manpower as well as of raw materials geared to France's industrial needs. It was in pursuit of this policy that France accorded such favourable treatment to Senegal and its citizens right from the start of her West African empire, but the actual material gain which she derived from this empire in the form of economic benefits was not so obvious.

A country which clearly derived immense material benefits from her colonial empire was Belgium. From the time that Stanley carved out the Congo basin for the King of Belgium until the country achieved its independence, Belgium obtained vast quantities of minerals and agricultural raw materials from the Congo.

After the initial economic and social developments which enabled the inhabitants of many of the colonial territories to become effective consumers of Western goods, colonial possessions with large populations became profitable markets for the manufactured products of the ruling nations. Generally speaking, the volume of trade conducted by the colonial powers with their colonies was small by comparison with trade conducted with other developed nations; but this trade was important nevertheless because it provided certain vital raw materials, such as rubber, cotton, cocoa and a number of minerals which were not available elsewhere. Moreover, because trade between the colonial powers and their colonies was based on the currency of the colonial powers, they were able to exercise a virtually monopolistic control over such trade, which in effect took the form of bilateral commodity exchanges in which the tune was almost invariably called by the colonial powers.

Finally, in almost all cases the colonies which had been acquired initially for their economic value had vital strategic advantages to offer; and this was an important consideration in a world where the security of trade routes and commerce was so dependent on military and naval

Fig. 8. British possessions in Africa at the end of the First World War. Note the continuous stretch of British-held territories from Cape Town to Cairo – the basis of the Cape to Cairo Railway dream

power. So important is this consideration that in many cases agreements between the colonial powers and their possessions regarding the maintenance of bases continued even after independence. In Africa, for example, thanks largely to the doctrines of imperial defence put forward by the late Field Marshal Smuts, Britain made considerable efforts during the last decades of the nineteenth century and the first three decades of the present century to establish a line of defences along the eastern plateau backbone of Africa; and although this dream, including the much discussed Cape to Cairo railway, was never wholly realised it formed for many years the basis of much of Britain's diplomatic effort in Southern and East Africa (Fig. 8).

From all this it can be seen that it is impossible to sustain wholly the claim that colonial possessions were nothing but liabilities for the colonial powers in Africa. Many of them offered very important advantages, although the full realisation of these advantages often took time. But the gain was not on one side only; in most instances the colonies also derived very considerable benefits from the colonial powers concerned. In particular, the association enabled important social, economic and cultural ties to be established between the colonies and the metropolitan countries, many of which have survived the subsequent attainment of political independence.

NEO-COLONIALISM

The ending of colonial rule in most countries in Africa has not, as was confidently assumed at the time of independence, resulted in the gaining of complete control by these countries over their economic or even their political affairs. Nominally, they are now sovereign states, but in reality most of them still remain under the economic and political control of their former rulers and the rich industrialised nations who direct the economic fortunes of the world. It is this kind of indirect or veiled control of the developing nations by the more advanced and powerful nations that is known as neo-colonialism. According to Nkrumah, 'the essence of neo-colonialism is that the State which is subject to it is, in theory, independent and has all the outward trappings of international sovereignty. In reality its economic system and thus its political policy is directed from outside' (Nkrumah 1965).

Most of the newly independent nations of Africa are deeply conscious of their economic backwardness and the appallingly low living standards of their people; and the desire to correct these deficiencies was among the principal motivating factors in their struggle for political independence. The argument was that without complete control over their political affairs they could not effectively control their own economic affairs and thus improve the economic lot of their people.

Subsequent experience has shown, however, that the achievement of political sovereignty does not automatically lead to economic independence because owing to the greatly superior economic and technological advantages which the developed nations enjoy they are still in a position to determine or even to dictate to a large extent the economic fortunes of the developing nations which depend on them for the very things, such as capital, manufactures, capital goods, technical know-how and entrepreneurial skills, which they need in order to modernise and upgrade their fragile economies. Many of the industrialised nations have exploited this advantage in every possible way and used it to gain indirect control over both the economic and political affairs of the developing nations, thus completely destroying the substance of their political sovereignty.

One area in which the developing nations suffer their greatest and most direct disadvantage is that of trade. It is only natural that in any trading arrangements between nations each party should seek to promote its own interests. However, because of the greatly superior advantages which they enjoy, the tendency has been for the rich industrial nations invariably to press their interests almost ruthlessly to the disadvantage of the poorer nations. In this sense neo-colonialism is not really a new phenomenon; it is new only in the sense that in its contemporary manifestation it has become greatly intensified in its effects by the sharp and ever-widening division of the world into two distinct groups made up of the very rich nations with rapidly growing economies, on the one hand, and the very poor nations, on the other hand, whose economies are either stagnant or growing at only a very slow rate.

The main problem facing the developing nations of Africa is how to break free from this economic bondage so that their recently achieved political independence can become a reality. It is within the means of both the rich and the poor nations to change the present state of affairs, and although morally the onus rests more on the richer nations than on the poorer ones, the responsibility is clearly a joint one. But recognising political realities of our world it is quite obvious that the initiative will have to come mainly from the developing nations, since it is they who are the sufferers. They must face up boldly to the realities of their plight and make determined efforts to correct the present imbalance by pursuing sound and realistic policies. It is futile to expect that their patterns of trade and economic development in relation to those of the rich nations can be changed overnight, but they can do a great deal to improve their economic lot by following a policy of self-reliance and mutual cooperation. The notion that the richer nations will help them out of their difficulties by the generous grant of aid without any regard for their own self-interest is a myth which they will do well to discard in spite of all the pious sentiments which are continually expressed about the need to bridge the gap between the rich and the poor nations.

If the developing nations plan carefully and follow sound economic policies they will succeed sooner or later in strengthening their economic position. Some of them could even succeed in becoming strikingly prosperous in view of their very rich natural resources. Complete economic independence, however, is a goal which neither they nor indeed any other country can hope to achieve in the circumstances of the modern world in which all countries are becoming increasingly interdependent, economically and politically. Herein lies the importance of cooperative arrangements at the regional and continental levels such as are currently being advocated by the Organisation of African Unity and the Economic Commission for Africa.

PART III

The states and their problems

5

North Africa

North Africa comprises Morocco, Algeria, Tunisia, Libya, Egypt, and the two small Spanish enclaves of Ceuta and Melilla. As defined by these political divisions the region covers an area of nearly 5.8 million sq. km (2.2 million sq. miles) and contains a population of some 79 million distributed as follows:

Political divisions	Area Sq. km	Sq. miles	Population 1975 or latest
Algeria	2,381,741	919,590	16,776,300
Egypt	1,001,449	386,660	37,233,000
Libya	1,759,540	679,358	2,444,000
Morocco	446,550	172,413	17,504,000
Tunisia	163,610	63,170	5,772,450
Total	5,752,890	2,221,191	79,729,750

Geographically, the region falls into four broad divisions: (a) the Mediterranean and Atlantic coastal fringes, (b) the Atlas mountain system, (c) the Nile valley, and (d) the desert (Fig. 9). Of these the desert with an area of approximately 1.8 million sq. miles is far the most extensive and is represented in each of the major political divisions, especially Libya, which is almost wholly desert, and Egypt, where the desert is relieved only by the 34,993 sq. km (13,500 sq. miles) of cultivated land formed by the valley and delta of the Nile. In Algeria, too, the desert occupies all except an area of 324,013 sq. km (125,000 sq. miles) in the northern mountainous and coastal parts; but Morocco and the greater part of Tunisia lie beyond its northern limits. The population of the region is concentrated in only about one-sixth of the total area, and scarcity of cultivable land, with all its attendant social consequences, is thus a serious problem throughout the region.

Culturally and politically, North Africa is one of the most complex regions in Africa. While the Moslem religion and the Arabic language which followed in its train have served over a long period of time to give the region a marked degree of cultural unity without a parallel in any part of the continent, this unity has been seriously undermined by a number of strong divisive forces resulting from historical, cultural

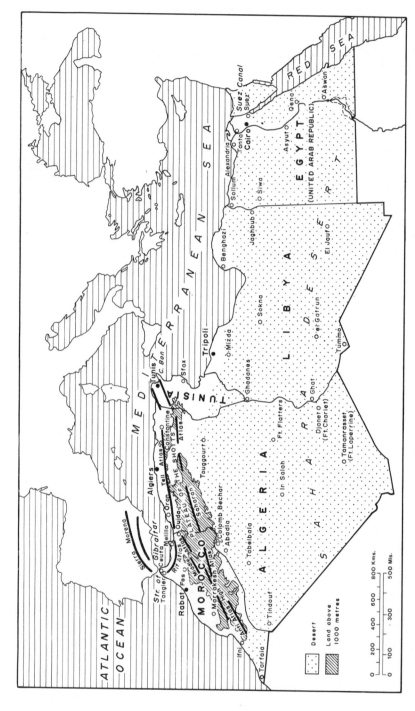

Fig. 9. Political divisions of North Africa, showing the distribution of coastal plains, mountains and desert

and geographical differences in the region as a whole and within its individual political divisions.

The first and most obvious line of division is that between the vast and sparsely populated desert lands in the south and the better watered and more densely populated Atlas and coastal areas to the north. Next and perhaps even more profound in its effects is the division between what is known as the Maghreb, that is, Morocco, Algeria and Tunisia, on the one hand, and Egypt, on the other, with Libya forming a kind of transition zone between the two.

The distinction between the Maghreb and the rest of the region has both a historical and geographical significance. The term *Maghreb* in Arabic means 'west' and was applied by the Arab conquerors of the seventh century to all those parts of their empire lying west of Egypt and the Arabian peninsula, which formed the true heartland of the empire. Three different 'Wests' were recognised, the Near West corresponding to present-day Libya and Tunisia, the Middle West, corresponding to Algeria, and the Far West, corresponding to Morocco and Mauritania. Today, however, the term is more often associated with Morocco, Algeria and Tunisia, the mountainous parts of which were known to the Arabs as *Djezira el Maghreb* (the Western Isle) (Hodder and Harris 1967. See also Harrison Church *et al.* 1964).

There is no doubt that the Maghreb, as defined above, bears a different cultural imprint from the lands to the east; for although both parts have been exposed for centuries to political and cultural influences from Europe and the Middle East, the effects of European colonisation appear to have left a much deeper impression on the countries of the Maghreb, just as those of the Arab connection are far stronger today in Egypt and Libya.

Similar contrasts can be found in the relations between these two major divisions of North Africa and the lands of Black Africa to the south of the Sahara. While the countries of Maghreb, especially Morocco and Algeria, have had long and active contacts with the Sudanic peoples of West Africa and today include some Negroid elements within their political boundaries, Egypt and Libya have remained virtually isolated from Black Africa on account of the very formidable barrier presented by the desert along their southern and western borders, although in the case of Egypt the effects of this isolation have been diminished by the tenuous access to the Black peoples of East Africa offered by the valley of the Upper Nile long before the era of modern land and air communications.

Despite its semblance of unity, the Maghreb itself is a land of striking internal contrasts. As a consequence of its long and varied history, in the course of which it has been subjected successively to invasion and colonisation by several outside powers, the region is marked today by a bewildering diversity of ethnic types. The indigenous inhabitants of the region were the Berbers, a term derived from the Latin word for barbarians. The Berbers are essentially a Caucasoid people with an occasional Negroid admixture who today, as in classical times, practise sedentary agriculture, although a few tribes, such as the Berber of the Middle Atlas, also practise seasonal transhumance. Through the years they have been subjected to successive waves of colonisation and outright conquest by other more powerful peoples from within the Mediterranean basin and the Arabian peninsula.

The earliest colonisers were the Phoenicians who established a number of settlements along the Mediterranean coast, most notably Carthage, which was located just north of the present site of Tunis. The Greeks also established a few coastal settlements, but it was the Romans who ultimately succeeded in effectively challenging the power of Carthage and in establishing an extensive empire in North Africa stretching from the Mediterranean to the Saharan border beyond the Atlas Mountains and from Egypt to present-day Morocco.

The Romans were succeeded by Arab conquerors in the seventh century who imposed their language as well as the Moslem religion upon the entire region and across the Strait of Gibraltar into the Iberian peninsula and southern France. The effects of this conquest have proved even more lasting than those of the Roman conquest, and despite the intervention of European colonial rule the Arabic language and the Moslem religion have persisted to this day and provided the region with its strongest unifying political bond. After the Arabs came the Turks, to be followed at a later date by the Spaniards, the Portuguese and the French.

Today, side by side with the Arab-speaking majority are minorities speaking Berber, French, Spanish and Italian, while in terms of race there are to be found Caucasoid (or white) groups, who form the overwhelming majority of the population, and scattered Negroid peoples who form the minority. In the sphere of religion, too, there are contrasts between the Mohammedan masses and the much smaller Jewish and Christian communities, while in the economic sphere equally striking contrasts exist between the sedentary cultivators of the more fertile coastal and intermontane areas and the nomadic and semi-nomadic pastoralists of the steppe and desert areas as well as between the relatively sophisticated and often more prosperous urban dwellers and the simpler and poorer folk of the rural areas.

All these contrasts have tended to provide fertile ground for internal disharmony. Especially serious has been the centuries-old hostility and distrust between the large Arab population and the Berber minority, who were displaced by the Arab conquerors and now occupy a subordinate social and economic position. Other causes of unrest have been the uneasy relations between the Arabs and the small Jewish populations for whom life has become even more difficult in recent years on account of the Arab–Israeli conflict and also the feeling of animosity between the Arab and the French population which were generated by the long drawn-out struggle for political emancipation from France.

Over and above the internal divisions and tensions that have persistently undermined social and political stability within the countries of the Maghreb, the region has had to contend, especially since the ending of European colonial rule, with the difficult problem of establishing its own identity unfettered by the deep-rooted ties which have developed with Europe over centuries of active political, economic and cultural intercourse and created an outlook which in certain important respects is distinctly more European than it is either Arab or African.

That this should be so is readily understood when one considers the geographical configuration of the western Mediterranean where the Maghreb is located. Here the pronounced northward thrust of the African continent between Cape Bon and the Strait of Gibraltar has resulted in bringing the countries of the Maghreb into very close physical proximity to the southern parts of Western Europe around the relatively small and virtually enclosed western basin of the Mediterranean to form a distinctive sub-region within the Mediterranean basin as a whole. In the course of the region's long history this western basin has exerted a strong attractive pull on all the surrounding lands and served as a ready means for invasion as well as for peaceful contact among them, while along the western rim of the basin, where Africa and Europe are barely separated from each other by the narrow Strait of Gibraltar, an easy land-bridge has been provided for contact between the Maghreb and the Iberian peninsula. In the light of these facts it is not surprising that the colonising powers that have been most active in the Maghreb in modern times have been Spain, France and Italy, all of which face Africa from the European side of the basin.

In contrast with the countries of the Maghreb, Egypt has enjoyed a relatively long period of internal cultural stability without any serious interference from the disruptive effects of prolonged or permanent European settlement. Like the countries of the Maghreb, it has also been subjected in the course of its long history to invasion and domination by numerous outside powers, notably, Persia, Macedonia, Rome, the Arabs, Turkey, France, and finally Britain, but it has succeeded throughout in retaining its cultural identity and the relatively homogeneous character of its racial composition, which ever since the Moslem invasion of the seventh century has remained almost wholly Arab in language, culture and outlook. But because of the strategic position it occupies at the eastern end of the Mediterranean and the fact that the Suez Canal lies within its territory it has been the focus of intense international interest and sometimes even rivalry, especially since the opening of the Suez Canal in 1869. In recent years this interest has deepened very considerably following the emergence of the Arab countries south of the canal as the world's most important source of oil, and the active role which Egypt has assumed since 1948 in the Arab struggle against the young Zionist state of Israel created in that year has placed it at the very centre of the highly explosive politics of the Middle East and the Arab countries of North Africa.

As a means of strengthening its position within this region, Egypt has been active in initiating a number of alliances since the end of the Second World War with its Arab neighbours. In 1958, under the inspiration of the late President Nasser, Egypt formed a Union with Syria under the name United Arab Republic. However, the Union failed to achieve anything concrete and was dissolved in 1961, when Syria withdrew from it, although Egypt retained the new name for itself. Then, in 1971 it joined with Syria and Libya to form a loose federation of Arab Republics, but again this failed to materialise. Since then various attempts at closer union between Egypt and Libya, including proposals for actual political amalgamation, have been made largely at the instance of Libya, but without any success. Despite these setbacks, Egypt has been able to bring the Arab countries much closer together than at any previous time in the region's modern history, using the common Arab hostility against Israel as the principal rallying point, and to strengthen its own position as the acknowledged spokesman for the Arab cause.

Although Egypt has no cultural or political commitments to either of the world's major blocs, its past political associations with Britain and France and the highly strategic position which it occupies in relation to the Arab oil-producing countries of the Middle East have given it very considerable advantages in its dealings with both the Western countries and the countries of the Soviet bloc. Furthermore, its active contacts with the Black African states south of the Sahara through the Organisation of African Unity have helped to forge valuable political links with these states and to range them on the side of the Arab countries in their recent struggle against Israel. Thus, although Egypt is not itself an important producer of oil, it has emerged in recent years as the main linch-pin in the oil politics of the Middle East and the principal point of contact between this explosive but vitally important region and the rest of the world.

In the context of African political affairs, Egypt, by reason of its geographical location at the north-eastern extremity of the African continent as well as on account of the policies it has pursued since the middle 1950s, has established itself as the natural bridge between

North Africa and the Middle East and between the Arab world and the black states of sub-Saharan Africa. As already pointed out, long before the development of modern communications across the Sahara Egypt enjoyed a measure of cultural and economic communication with Black Africa, especially East Africa, by means of the Nile waterway. However, the main focus of its political interest remained North Africa and the Middle East.

A new phase in the relations between Egypt and Black Africa began in 1956 as a result of Egypt's active participation in the Bandung Conference of Afro-Asian states, which among other things expressed its solidarity with the various nationalist movements struggling for the political emancipation of colonial Africa from European rule. Thereafter, the country's president, Abdel Nasser, working closely with other African leaders, such as Kwame Nkrumah of Ghana, established himself as one of the continent's most radical leaders and staunchest opponents of colonialism.

As regards purely Arab affairs also, Nasser became a martyr in the eyes of the Arab world and a centre of attention following the abortive attack against Egypt launched between October and November 1956 by Israel, Britain and France but subsequently halted upon the insistence of the United States and the Soviet Union. Before then the decision of the United States and Britain to withdraw their financial support for the construction of the Aswan High Dam upon which Egypt had pinned its hopes for future economic development had given Nasser a convenient pretext for nationalising the Suez Canal which had been mainly under foreign ownership since its opening in 1869. The nationalisation of the canal particularly angered Britain and France because of their considerable financial and strategic interests in it and was no doubt chiefly responsible for their decision to invade Egypt, but Nasser's very bold action and the subsequent humiliation of Britain and France served to give his prestige a very big boost in the Arab world and in colonial Africa.

Since then, Egypt has become the front-line Arab state in the protracted conflict between Israel and the Arab world. Despite considerable financial and material support from some of the richer Arab states, especially Saudi Arabia, the country has continued to bear the main brunt of Israel's very considerable military strength and also to receive most of the credit for the Arab success in resisting attacks from Israel.

With Nasser's death in 1970, Egypt's role in African affairs, as distinct from the affairs of North Africa, has dwindled quite considerably, but there has been a corresponding rise in the country's stature in the Arab world. There is no doubt that although Egypt's economic resources are quite limited by comparison with some of its neighbours and in relation to its population of over 37 million, it is the most economically and industrially developed of the Arab states of North Africa and the Middle East, with a large reservoir of highly trained manpower. It also has the largest and best-equipped army in the entire region and, what is more, is now the sole master of the Suez Canal, which is a sea lane of crucial importance not only for most of the world's richest and most powerful nations but for international trade generally. In purely strategic terms, few countries in the world – and certainly none in Africa – enjoy such an enviable position in relation to a major seaway. Egypt is fully conscious of this fact and in recent years, under Nasser's successor, President Sadat, has displayed remarkable skill in exploiting this unique advantage for securing financial assistance from the two major opposing power blocs for bolstering up its economy and for building up its military strength.

The vast area of desert along the Mediterranean coast between Egypt and the Maghreb is occupied by Libya, which is essentially a geographical as well as a political transition zone between the true Arab heartland to the east and the Maghreb countries with their strong cultural attachments to Europe to the west. Being far less committed culturally and politically to Europe than any of the other North African countries and not being in direct physical contact with Israel, as is the

Crude oil production in North Africa and the Middle East: 1970 and 1974

Country	Production (in thousands of 42-gallon barrels)	
	1970	*1974*
Algeria	371,767	372,753
Libya	1,209,314	555,291
Morocco	335	191
Tunisia	34,296	31,841
Egypt (United Arab Republic)	119,165	53,715
Bahrain	27,973	24,597
Iran	1,397,460	2,210,627
Iraq	567,726	679,803
Israel	31,798	36,500
Kuwait	1,090,040	830,580
Oman	212,210	106,046
Qatar	132,456	189,348
Saudi Arabia	1,387,266	2,996,543
Syria	29,356	45,352
United Arab Emirates (formerly Trucial States)		616,485
Abu Dhabi	252,179	
Dubai	31,321	
Total	6,894,662	8,749,672
World output	16,689,617	20,518,139

(Figures taken from the 1972 and 1976 World Almanac, Pittsburgh Press, New York)

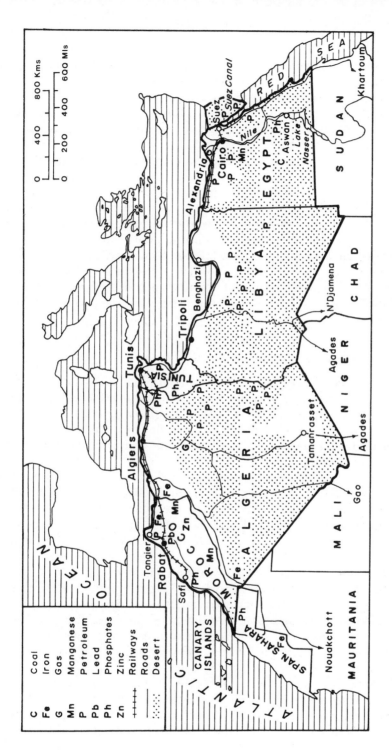

Fig. 10. North Africa: minerals and communications

case with Egypt, Libya has been able, despite its militancy, to follow a far more independent line in the struggle against Israel than any of its Arab neighbours. In this it has been assisted very considerably by its immense oil wealth, which has enabled it to back its policies with impressive material support (Fig. 10). But this very fact has tended to make Libya highly intolerant of outside opposition, with the result that it has found cooperation difficult with its less rich but more politically mature Arab neighbours whose support is vital for the achievement of the internal political and cultural stability which the country desperately needs. It is, however, a country of great strategic importance on account of its oil resources which, as shown in the table on page 90, far exceed those of any other country in North Africa and placed it third in 1970 and sixth in 1974 among the oil-producing countries of the Arab world. Much of the influence which Libya now wields in the Arab world is owed to this single resource, and there is no doubt that provided its use is backed by wise policies the country's political role not only in the Middle East and North Africa but in the world as a whole could be greatly enhanced.

MODERN EVOLUTION OF NORTH AFRICAN STATES

With the exception of Egypt, whose foundation as a distinctive state goes very much farther back in history, none of the present-day states of North Africa appears to have enjoyed anything beyond a very vague existence until during the period of Turkish rule in the area, approximately between the middle of the sixteenth and the beginning of the nineteenth century.

Under the Turks hereditary dynasties ruled by Beys were founded in Tripoli and Tunis, while Algiers became an oligarchic republic ruled by a Dey elected for life, thus laying the foundations of the present political divisions of Libya, Tunisia and Algeria. Turkish rule did not, however, extend to Morocco, which more or less retained the boundaries of the ancient Roman province, *Mauritania Tingitana*. None of these units had precisely drawn boundaries in the modern sense until the arrival of the European colonising powers of modern times. But it must be admitted that even under these powers boundary delimitation and demarcation were not easy, partly as a result of the problems presented by the desert terrain and partly because of the abundance of intractable border disputes among the local people themselves.

The first European powers to establish footholds in North Africa in modern times were Spain and Portugal. As far back as 1415, even before the final defeat of the Moslem invaders of the Iberian peninsula at Granada in 1492, the Portuguese had taken control of Ceuta, which lay within Moroccan territory. This was followed early in the following century by the establishment of a string of strongholds extending along the Atlantic coast of Morocco from Tangier to Agadir. At the same time the Spanish also seized Melilla, Oran, Algiers, Bougie, Tunis and Tripoli along the Mediterranean coast. However, because of the preoccupation of these two powers with more profitable overseas ventures in other parts of the world, most of these strongholds were lost to the local people, with the exception of Mazagan, which remained in Portuguese hands until 1790, Ceuta, which was ceded to Spain by Portugal in 1668, and Melilla, which was retained by Spain and still remains under Spanish control.

For several years, apart from the footholds established by Spain and Portugal, the European powers seem to have conceded the right of the North African countries to exist as autonomous regions under their own local rulers without outside interference and were content with merely establishing diplomatic and trading relations with them, although they took periodic action to check the activities of the Barbary pirates who menaced shipping in the Mediterranean from their coast. However, in 1830, France which had more or less succeeded Spain and Portugal as the most influential power in the Maghreb found an excuse, following a quarrel with the ruling Dey, for embarking on the military conquest of Algeria, mainly in order to gratify the French army which had been demoralised by its setbacks in Europe under Napoleon. The conquest proved more difficult than had been expected and lasted until 1884 on account of the strong opposition which the French encountered from the Arab and Berber populations, especially the latter.

An important aspect of French rule in Algeria and one which was to have far-reaching consequences was that although the French embarked on their conquest with the declared aim of freeing the country from oppression and of giving the people the benefits of French civilisation, no sooner had the conquest been completed than they began the large-scale settlement of European colonists, many of them drawn from Spain, Italy and other non-French parts of Europe, and the massive and ruthless expropriation of land from the native population, amounting by the early twentieth century to about 40 per cent of the country's entire cultivated area. Just before the country attained its independence in 1962 there were as many as 400,000 French settlers or 'colons' together with an equal number of non-French European settlers enjoying French citizenship.

From Algeria the French moved on to Tunisia, which was occupied without much difficulty in 1881, largely through diplomatic agreements with the local ruler or Bey and with the backing of Britain which was prepared to trade off its interests in the area in return for a free hand in Egypt.

Here also the French pursued the policy of settling European colonists, but on a much smaller scale than in Algeria; hardly any land was expropriated and only 20 per cent of the cultivated land was taken over by them. Another important departure from their Algerian policy was that although effective political power passed into French hands the local Dey was allowed to retain nominal authority and the country was given the status of a protectorate.

The next country to engage the attention of the French was Morocco, whose attractive economic possibilities and strategic location at the western entrance to the Mediterranean made it a particularly coveted prize among the European maritime powers with interests in the Mediterranean, notably Britain, Italy, Spain, France and Germany. However, by agreeing to Italian control of Libya, British control of Egypt, German control of the Cameroons, and Spanish control of the northern tenth of the country, France was allowed to assume control of the country, which thus became a French protectorate in 1912. A further important development occurred in 1923 when Tangier and its surrounding territory which lay within the Spanish zone was turned into an international zone under French, Spanish and British control.

French pacification of Morocco was much more gradual and humane than in any other part of North Africa both because of the strength of local opposition and because of the high regard in which the first French Resident General, Marshal Lyautey, held the rich local culture, and it was not until well after the First World War that the pacification was effectively completed. Even so the Saharan borderlands continued to defy administration and remained a difficult frontier zone throughout the period of French rule. Just as they did in Algeria and Tunisia, the French settled colonists in Morocco, but on a much smaller scale, only 8 per cent of the cultivated land being occupied by them. At the time of independence in 1956 there were about 360,000 foreigners in the country, of whom 84 per cent were French and the rest Spaniards living in the northern Spanish zone.

The last North African country to fall into European hands was Libya, which is the second largest in size though the least populous of the countries within the region. Although it shares many common cultural characteristics with the countries of the Maghreb, its geography and history clearly set it apart from them and it is best regarded as a transition zone between these countries and Egypt, which belongs more truly to the Moslem heartland of the Arabian peninsula. Libya's long, featureless coastline has almost desertic conditions and is effectively separated into two relatively small but distinct enclaves of somewhat moister and more habitable land by a wide salient of some 480 km (300 miles) around the Gulf of Syrte, where the desert reaches right down to the coast. There can be little doubt that it was because of its inhospitable environment that Libya attracted so little attention from the major colonial powers in the nineteenth century.

Like other parts of North Africa Libya has experienced domination in the past by several external powers, such as the Egyptians, Phoenicians, Greeks, Romans, Vandals, Byzantines, Arabs, and in more modern times by the Spaniards, Turks and Italians, all of whom have left their imprint on the country.

To the Greeks the term Libya generally denoted the whole of the North African coast between the Nile delta and the Atlantic; but later it acquired greater precision and under Turkish rule between the sixteenth and nineteenth centuries it came to be applied to the area within the country's present boundaries, comprising the three provinces of Cyrenaica and Tripolitania along the eastern and western extremities of the coast and Fezzan in the desert interior. Following the Italo-Turkish war of 1911–12 direct rule by Turkey came to an end and the country seemed to be on the verge of independence. However, Italy with the tacit support of the other European powers with territorial interests in the Mediterranean continued the fight against the local population until it gained full control of the country in 1914. Although the Italians suffered a temporary setback during the First World War when the Tripolitanians and the Senussi tribesmen of Cyrenaica drove them back to the coastal cities, they were able to regain full control after the war through a series of brutal campaigns and political measures between 1921 and 1932 and to unite the three provinces into a single colony.

Under Mussolini the Italians sought to revive the past glories of the Roman Empire by embarking on several grandiose but brilliant schemes for the development of cities, roads and water supplies, many of which proved later to have been economically misconceived. As a means of relieving acute overpopulation and unemployment problems at home, the Italians settled large numbers of peasants in the more hospitable areas around Tripoli and in Cyrenaica, while the pastoral elements of the Moslem population were pushed into the less developed desert fringes, and, by 1939, of the 300,000 Italian settlers proposed as the ultimate target by Mussolini as many as 110,000 were already in the country.

In 1939 Italy's annexation of the territory was taken a step further by its incorporation into the metropolitan kingdom of Italy. Throughout, however, the only real beneficiaries of the Italian presence in Libya were the colonists, the Moslem population being largely neglected or mercilessly exploited. But Italy's entry into the Second World War on the side of the Axis powers brought her Libyan colonial adventure to an abrupt end and by 1943 the Western Allies had taken control of the territory and practically the entire Italian population had been evacuated. The administration of Cyrenaica and Tripolitania was taken over by the British, while the French assumed responsibility for Fezzan. At the end of the war the future of Libya was referred to the United Nations, which decided to grant the country full independence under a federal constitution as soon as possible, and on 24 December 1951 the new United Kingdom of Libya was officially proclaimed. But since Italy, the country's real mentor, had by then been removed from the African colonial scene it fell to Britain and the United States to give Libya most of the initial financial and technical support necessary to enable it to find its feet as an independent nation.

Thus, although Libya was the last of the North African countries to come under colonial domination, it was the first, apart from Egypt which became independent as far back as 1912, to achieve its independence after the Second World War. This was not because it was the most developed or politically mature but because by a pure accident of history its administration came under the United Nations immediately after the war and the country therefore reaped immediate benefits from the organisation's pledge to promote the political emancipation of colonial subjects. But a consequence of this rapid transition from colonial status to full independence was that unlike the other countries in the Maghreb, Libya never developed the strong cultural links with any of the European powers which have proved to be a source both of strength and of weakness in the case of these countries.

THE POST-WAR INDEPENDENCE MOVEMENT AND SUBSEQUENT DEVELOPMENTS

North Africa was the first part of the African continent to feel the effects of the post-war independence movement. As already stated, Libya won its independence in 1951. In 1956 Morocco and Tunisia gained their independence from France in rapid succession (on 2 and 20 March, respectively) without any serious difficulty. The Spanish zone of Morocco as well as Tangier were also turned over to Moroccan control in the same year, but Spain retained the fortified cities of Ceuta and Melilla. Another possession which Spain retained in Morocco was the enclave of Ifni, but this was ceded to Morocco in January 1969. Algeria's progress towards independence, however, took a different course and it was not until 1962 that the country was able to win its freedom from France, and that only after a long and bitter struggle that lasted eight years and in the process nearly embroiled France itself in civil war owing to the stubborn opposition put up by the 1 million or so French settlers with the backing of the Algerian army, which right from the start of French rule in the country had been noted for its independent stand in political matters. The French settlers regarded

Algeria as their own rightful possession and an extension of metropolitan France which should on no account be handed over to the
indigenous Moslem population.

With Algeria's attainment of independence all the countries of North
Africa became politically free, and the strong cultural ties which had
existed between them and the countries of Western Europe, especially
France, began to be undermined by a rapid reassertion of their historic
ties with the rest of the Arab world. To the Moslem religion and the
Arabic language which had initially provided the main bond among
them was added a new sense of unity born out of the common hostility
among the Arabs towards the newly established Zionist state of Israel.
Thanks to this hostility, the North African countries, especially Egypt
and Libya, have been drawn more closely than ever before into the
highly complex and explosive politics of the Middle East.

Always a troubled and unstable part of the world, the Middle East
has become even more unstable since the end of the Second World War
on account of the conflict between the Arabs and Israel, while the
importance of the region as a source of oil has made its internal politics
a matter of deep international concern. The Arabs have been quick to
recognise the potency of their oil wealth as a political and economic
weapon for compelling the rest of the world to give them the support
and sympathy to which they feel themselves entitled in their struggle
against Israel, as was so forcefully demonstrated in 1973–74 by their
embargo on the shipment of oil. But by using this weapon the Arabs
have heightened the tempo of international tension and rivalry in the
Middle East and made its internal affairs the common concern of the
whole world. These developments have inevitably affected North
Africa because of its involvements with the Middle East, and the region
has accordingly lost any claims it might once have had to be regarded
and treated as a politically uncommitted part of the world in relation to
the great powers.

TERRITORIAL AND STRATEGIC ASPECTS

Until the discovery of vast reserves of oil and natural gas in Algeria and
Libya between 1956 and 1961 North Africa was generally a poor region
in terms of natural resources, and agriculture was the principal
economic activity. The only countries where mineral production was
important were Morocco and Tunisia. Morocco was a leading world
producer of phosphate, but its mineral exports also included iron, coal,
manganese, cobalt and zinc, while Tunisia produced iron, phosphate,
lead and zinc (Fig. 10). In the Maghreb countries the settlement of
large numbers of European colonists, especially in Algeria, gave
agriculture an important boost, but nowhere did this result in the

development of a really prosperous economy. The least developed country of all was Libya, where the extent of desert rendered cultivation well nigh impossible except in a few limited areas in Cyrenaica and around Tripoli along the coast. Egypt, too, was not particularly well endowed, although the valley and delta of the Nile provided a favourable environment for both subsistence and commercial agriculture.

Throughout the period of European rule, the principal value of North Africa lay in its strategic location along the southern Mediterranean and the opportunities it afforded for the establishment of naval and military bases from which the vital sea routes through the Mediterranean could be controlled and safeguarded. France held important air and naval bases at Port Lyautey (now Kenitra) on the Atlantic coast of Morocco and at Oran on the Mediterranean, as well as at Algiers in Algeria and at Bizerta in Tunisia, while Italy held Tripoli and Tobruk in Libya, and Britain, Alexandria and Port Said at the entrance to the Suez Canal under special treaty arrangements with Egypt.

From earliest times the outlook of North Africa had been towards the Mediterranean, of whose civilisation it formed a part, but the European control of the region further accentuated this orientation. Apart from the European air and naval bases which had coastal locations, the capital cities of all the North African countries were located either on the sea or near by. Thus Morocco's capital was at Rabat, only a short distance from the Atlantic coast; Algeria's capital was at Algiers, which was an important naval base on the Mediterranean; the capital of Tunisia was Tunis, also on the Mediterranean, while Libya's two capitals, Tripoli (now Tarabulus) and Benghazi, were both on the Mediterranean, and Cairo, the capital of Egypt, stood at the apex of the Mediterranean coast.

Except during the brief period of Carthaginian ascendancy, none of the North African countries has ever emerged as a major naval power within the Mediterranean, and any influence which the region has exercised in military and naval affairs during modern times has been essentially a reflection of the strength and interests of the foreign powers that ruled the area. There is no better illustration of this than the role which North Africa played during the Second World War, when the region became one of the major theatres of the war and the scene of a number of decisive battles between the Allied forces and the Axis powers. Apart from the unparalleled advantages which the control of the region offered for the protection of shipping through the two vital gates of the Mediterranean, namely, the Strait of Gibraltar and the Suez Canal, and within the sea itself, it enabled the Allies to outflank the Axis powers who at the time were in control of practically the whole of Europe and parts of North Africa and to deliver vital blows against them along their most vulnerable flank.

In previous centuries the Mediterranean had been essentially a European lake and at the height of the colonial era it was dominated almost exclusively by the powerful British navy. The Second World War led for the first time in history to the establishment of a powerful United States naval presence within the Mediterranean; and since the war the United States sixth fleet has remained in the area as a counterpoise to the Soviet Union which has direct sea access to the Mediterranean from its Black Sea ports and whose influence in the Middle East has grown steadily ever since the war. Potentially, North Africa still occupies an important strategic position in relation to the Mediterranean, but in strictly military terms the powers that really control the sea and its adjoining areas are the two super-powers, America and Russia. And the areas to which their attention is particularly directed are the Arab countries of the Middle East and North Africa from which the major part of the world's oil supplies is derived.

In the Suez Canal, however, the region – and more specifically Egypt – possesses a strategic asset of timeless value, whose importance has been greatly enhanced since the Second World War with the emergence of the Middle East as the world's single largest source of petroleum. But although the canal runs through Egyptian territory, from the time of its opening in 1869 until its seizure and nationalisation by President Nasser in 1956 it was owned and controlled by an international company in which Egypt held only a minority share. The strategic importance of this 161 km (100 mile) waterway was clearly underlined at the time by the angry reaction of the Western powers to Nasser's action. After a futile attempt by Britain, France and Israel to assert their rights to the use of the canal by the use of force the Western powers were obliged to resort to negotiations and diplomatic overtures in order to gain accommodation with regard to the use of the canal. But its control has been in Egyptian hands ever since.

In 1949, after the first Arab–Israeli war, Egypt had given a forceful demonstration of the strategic importance of the canal by denying its use to Israel and to all ships trading with Israel. But an even more forceful demonstration of the Canal's importance and the power which Egypt now wields through its control occurred in June 1967 when, following the war with Israel, Egypt closed the canal to all shipping for the following eight years. This action forced all countries that had previously relied on the canal for access to the Middle East and the Indian Ocean to use the much longer and in some cases infinitely more expensive route around the Cape of Good Hope, which adds as much as 7,730 km (4,800 miles) to the sea journey between London and Kuwait and 5,800 km (3,600 miles) to that between New York and Saudi Arabia.

In North Africa itself the post-war era has seen a marked dwindling of Western influence owing to the general failure of the West to support the Arabs in their fight against Israel in contrast to the Soviet Union which has openly declared its support for the Arab cause. As a result most of the military and naval bases previously controlled by the West within the region have had to be abandoned. Nevertheless, because of the region's strategic importance the Western nations, especially France, have made and continue to make desperate efforts to maintain their links with it through diplomatic contacts and the conclusion of trade and military agreements.

BOUNDARY PROBLEMS

As has already been noted, the delimitation of precise boundaries in North Africa, particularly in the desert areas, has never been easy in view of the difficult physical problems posed by the environment. However, during the colonial period there were few serious eruptions of boundary problems because the people likely to be involved in such disputes were either under the same European power or under powers generally friendly towards each other. With the ending of colonial rule many of these problems have assumed a new importance and led to a number of serious conflicts between some of the countries in the region.

The most serious problems have occurred in the Maghreb, where disputes have broken out between Morocco and Algeria, and between Morocco, Mauritania and the Spanish territories adjacent to Morocco. The dispute between Algeria and Morocco has centred on a stretch of about 644 km (400 miles) along Morocco's south-eastern border which was never precisely demarcated by the French but which has now acquired special interest on account of the discovery of extensive iron ore deposits in the region of Gara Djebilet, near Tindouf, a short distance inside Algeria's present boundary as defined by the French at the time of the country's independence. The Moroccans claim that the area was historically always a part of their country and that its inclusion within Algeria was part of a French design to hem Morocco into the smallest possible area at the time of the French conquest of the country in 1912 so as to extend the area under direct French control in Algeria.

In the western Sahara, too, Morocco has tried to lay claim to territory at present under Algerian control, again on historical grounds; but it is suspected that the real reason behind this claim is that although the area in question was formerly considered as useless desert it has recently acquired considerable economic importance on account of the discovery there of valuable mineral resources.

Along its southern border also, Morocco has laid claim to the whole of Mauritania and Spanish Sahara. Before Mauritania became independent in 1960 the Mauritanians themselves, concerned about the very limited economic base of their country, seemed anxious to effect a political merger with Morocco; but with the subsequent discovery of rich iron ore deposits at Fort Gouraud in Mauritania and the greatly improved economic prospects which this held out to the country enthusiasm for the merger died down. In the light of the changed situation Morocco decided on the extraordinary step of claiming the whole of Mauritania as part of its rightful territorial inheritance.

It is quite unlikely that this claim will succeed, and indeed it seems to have been dropped for the time being. However, in the case of the former Spanish Sahara (now known as the Western Sahara) both Morocco and Mauritania have moved in since the withdrawal of Spain at the end of 1975 and in the face of strong opposition from the inhabitants partitioned the country between themselves on the grounds that it formed part of their territories before the European colonisation of the continent. Six years earlier in 1969 Morocco had succeeded in regaining ownership of the small Spanish enclave of Ifni located on its Atlantic coast, but two other enclaves, Ceuta and Melilla on the Mediterranean coast of Morocco still remain in Spanish hands.

In the eastern part of the Maghreb the boundaries drawn up during the period of colonial rule have proved far more stable, although here too Tunisia has demanded, largely on economic grounds, that both Algeria and Libya, which have vast expanses of territory at their disposal, should concede part of their territories to it in order to enable it to obtain a share of the underground water and mineral resources of the Sahara which it needs vitally for its economic viability.

That none of these potentially dangerous boundary claims has resulted in open conflict between the countries concerned has been due mainly to the efforts of the Organisation of African Unity and its influence in promoting generally harmonious relations among its members. However, the very fact that such problems should arise within a region whose various governments have over a long period of time consistently acknowledged the need for closer political cooperation based on their common cultural origins and their common political aspirations serves to underline the deep-seated nature of the conflict which still exists in the region between the claims of narrow national interests and the broader objectives of regional political unity.

6

West Africa

West Africa occupies a unique position in the political geography of Africa. It is and, as far as the available evidence suggests, has always been predominantly an area of black or dark-skinned people without any significant European settler populations. Indeed, the term *Sudan*, which is applied to the sub-Saharan zone along its northern borders, is derived from the Arabic term *Bilad-es-Sudan*, meaning 'the land of the Blacks', while the word *Guinea*, which is more often applied to the southern and coastal parts of the region, appears to be the Moroccan Berber equivalent, *Akal-n-iguinawen*, a term whose usage dates back to the twelfth century (Fage 1961.)

After establishing their first footholds in North Africa it was to West Africa that Portugal and Spain turned their attention in their search for a sea route to India and the Far East, to be followed soon afterwards by the other maritime nations of Europe. It was also in West Africa that the independence movement of Black Africa, which followed the Second World War, first got under way after nearly five centuries of colonial rule in some cases. Unlike the rest of the continent, especially East, Southern and North Africa, where the climatic and other physical conditions made permanent white settlement possible, West Africa never experienced permanent settlement by Europeans, and trade and other forms of economic exploitation remained the primary object of European colonising effort within the region.

The outstanding physical characteristics of West Africa are its very large area, its compactness and its tropical climate, all of which have had a profound effect on its political geography. But perhaps its most striking and distinctive physical characteristic is the prevalence of low altitudes. Except for a few scattered and lofty mountain masses and peaks, notably the Mount Cameroon, the Bamenda and Adamawa Highlands, the Jos Plateau, the Guinea Highlands, the Fouta Djallon Mountains and the Air Massif, practically the entire area consists of extensive and greatly eroded plateau surfaces composed of very ancient rocks standing at altitudes of less than 1,000 m (3,000 ft) and extending almost right down to the sea, leaving only a few narrow coastal plains.

In all these respects West Africa represents a completely different world from North Africa, although the two regions share a common physical environment in their Saharan borderlands where their present-day political frontiers meet and where before the European colonisation of the Guinea coast they enjoyed relatively close and regular economic and cultural contacts with each other by means of the caravan routes that linked them across the Sahara. But while North Africa clearly belongs to the Mediterranean world and the Middle East and is inhabited predominantly by light-skinned or Caucasoid peoples, West Africa is essentially tropical in character and is inhabited principally by black peoples of Negro stock whose outlook is wholly and unmistakably African.

DEFINITION OF WEST AFRICA

Although the core of the region has never been in dispute, there has always been a considerable divergence of opinion among geographers as to the exact definition of its outer limits; for while the western and southern boundaries which are formed by the Atlantic Ocean are clearly marked, the landward boundaries along the north and east are much less easy to define with any degree of precision. On purely geographical grounds the southern edge of the Sahara which marks the limit of effective human settlement would seem to form the most obvious boundary along the north, while along the east the Adamawa Highlands which mark the approximate eastern limit of the area subject to the full climatic effects of the monsoon and harmattan winds so typical of West Africa would seem to form the most obvious and convenient boundary.

Political divisions	Area		Population 1975 or latest
	Sq. km	Sq. miles	
Benin	112,622	43,483	3,112,000
Cape Verde	4,033	1,557	294,132
Gambia	11,295	4,361	523,716
Ghana	238,537	92,100	9,866,000
Guinea	245,857	94,925	4,416,000
Guinea Bissau	36,125	13,948	528,000
Ivory Coast	322,463	124,503	4,885,000
Liberia	111,369	43,000	1,708,000
Mali	1,240,000	478,764	5,697,000
Mauritania	1,030,700	397,953	1,697,000
Niger	1,267,000	489,189	4,599,785
Nigeria	923,768	356,669	62,925,000
Senegal	196,192	75,750	4,136,264
Sierra Leone	71,740	27,699	2,983,000
Spanish Sahara	265,898	102,663	76,425
Togo	56,000	21,622	2,222,000
Upper Volta	274,200	105,869	6,032,000
Total	6,407,799	2,474,055	115,223,322

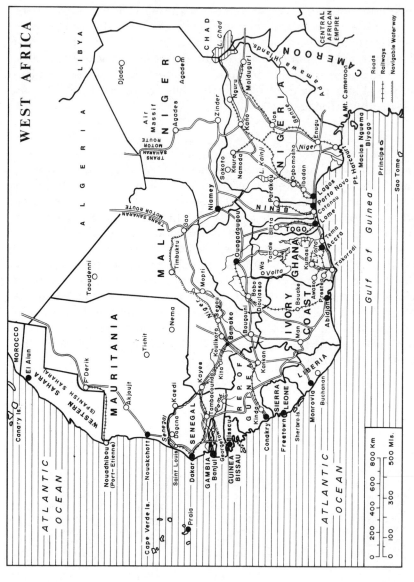

Fig. 11. West Africa: political divisions and communications

From the point of view of political geography, however, the region is best defined as the whole of the western bulge of Africa lying south of Libya and the Maghreb states of North Africa and west of the Federal Republic of Cameroon (Fig. 11). Thus defined the region covers an area of 6,407,799 sq. km (2,474,055 sq. miles) and, according to the most recent estimates, contains a population of over 115 million distributed as shown in the table on page 103. Despite its great size, it is highly compact in shape, measuring approximately 3,622 km (2,250 miles) from east to west at its widest point and 3,059 km (1,900 miles) from north to south.

West Africa is politically the most highly segmented region in Africa; it contains no less than seventeen of the continent's fifty-one political divisions and has a total of some 37,030 km (23,000 miles) of international boundaries. This is a direct result of the intense competition for trade and territorial control which took place among the European powers between the fifteenth and the twentieth centuries, especially along the limited stretch of coast known as the Gold Coast. During this period the European nations that acquired possessions or established trading posts in the region at one time or another included Portugal, Spain, England, France, Denmark, Sweden, Brandenburg, Holland and Germany (Fig. 7).

POLITICAL EVOLUTION

Without any doubt the most powerful factor in the modern political evolution of West Africa has been the European colonisation of the region. Although the initial purpose of the voyages of discovery which first brought the Portuguese to the Guinea coast was the search for an alternative sea route to India and the Orient which would circumvent the Arab menace which blocked access to the lucrative trade of the area via the eastern Mediterranean, the discovery of rich and abundant supplies of gold and spices along the Guinea coast soon led to an active trade and attracted other European nations into the region. With the discovery of America and the development of a new trade in slaves for supplying cheap labour to the American colonies West Africa assumed an even greater commercial importance for Europe.

Before the European arrival the principal centres of trade and population had lain in the Sudan and Sahelian zones along the Saharan border. As the new trade with Europe along the coast grew in importance, however, a steady movement of peoples began towards the coast and a series of new and powerful kingdoms emerged within the Guinea and coastal zones. Exactly how many such kingdoms there were is difficult to say, but most of them were quite small in size and mainly conformed with the existing linguistic and tribal groupings. But they were sovereign and autonomous entities with all the essential attributes of complete or embryonic nation states.

As has already been pointed out in Chapter 4, the European colonisation of West Africa was by no means an easy process. Apart from the difficulties presented by the climatic conditions of the Guinea coast, which was the main focus of European interest, as well as by the generally impenetrable character of the equatorial and tropical forests along the coast, the lack of suitable inlets and the rifeness of deadly environmental diseases, especially malaria, yellow fever and dysentery, the local inhabitants proved extremely hostile to the establishment of permanent settlements by the European traders.

Despite these difficulties, there were other factors which favoured the establishment of trading posts. The first was the proximity of the region to Europe and America and the location of all three regions along the shores of the North Atlantic. This enabled the European traders to reach West Africa quite quickly and, when necessary, to bring ready relief to their beleaguered forts. It also made it easy, when the slave trade developed, for the triangular trade to be established whereby various manufactured goods were shipped from Europe to West Africa in exchange for slaves who were in turn shipped across the Atlantic to America, from where exotic raw materials like cotton and cane sugar produced by slave labour were exported to Europe. Also favourable were the winds and ocean currents off the coasts of West Africa which greatly facilitated the movements of the sailing ships then in use in their journeys to and from West Africa as well as from West Africa to North America and the West Indies at the right times of the year (see Chapter 4).

For the first century after their arrival at Elmina the Portuguese held a monopoly of the West African trade, although in the initial period they were in competition with Spain until by the Treaty of Tordesillas in 1494 the Pope assigned all lands discovered approximately east of longitude 60°W to Portugal and those to the west to Spain, thus effectively removing the latter from West Africa except for lands they had already acquired. But the Portuguese monopoly began to wane in 1598 when the Dutch, then in revolt against Spain, began to make serious inroads into West Africa and to establish trading posts.

By 1642 all the Portuguese forts on the Gold Coast had fallen to the Dutch, who needed slaves to develop their newly acquired empire in America where they had also been active. From then until the early years of the eighteenth century West Africa, especially the Gold Coast, became the scene of intense competition among the maritime nations of Western Europe, and English, Danish and, for a brief period, Swedish and Brandenburger forts appeared, not to mention the claims staked by various interlopers from these and other nations who were attracted to the region by the slave trade. But throughout, the really effective domain of these nations remained the high seas and the forts they occupied on land leased in most cases from the local inhabitants. The rest of the land remained fully in the hands of the people of West Africa themselves.

We know very little about the internal situation of West Africa during this period, except for the immediate coastal areas which came under the direct observation of the European traders. The large empires such as Ghana, Mali and Songhai which had arisen in the western and west-central Sudan from the fourth century AD onwards had mostly disintegrated and had been succeeded by more modern states such as Bambara, Hausaland and Bornu. But in the southern part of the region the people lived mostly in small tribal units desperately trying to establish themselves in the face of the tremendous shock waves set in motion by the slave trade, although here also, as has already been mentioned, a number of major groupings and kingdoms were beginning to emerge.

For the next four centuries or so after the establishment of the first European settlements along the coast, that is, until the last quarter of the nineteenth century, the indigenous pattern of states remained very fluid. But, except along the coast where the European trading posts sometimes influenced the course of events, such changes as occurred were almost entirely the result of internal factors.

During this period relations between the local states were often strained, and inter-tribal wars were quite common, a fact which no doubt served the interests of the European traders whose main concern was the securing of slaves. Owing to the general insecurity of the period it was quite common for states lying far distant from a powerful chief to attach themselves to his state as a means of ensuring protection against attack from hostile neighbours. The net result of this was the eventual development of a somewhat complicated pattern of states, several of which had 'islands' or exclaves lying inside the territories of other states, as can be seen in present-day Ghana. Few states had properly defined boundaries; instead, they were separated from each other by vague frontier zones that oscillated with the changing fortunes of war. Thus, even today many of the boundaries between the traditional states are still the subject of endless disputes, some of which date back hundreds of years. However, where convenient natural features were available they were often employed for the demarcation of boundaries.

With the formal abolition of the slave trade in 1807, many of the European nations lost interest in West Africa and either abandoned or sold their trading posts. As far as the Gold Coast was concerned the Portuguese had already been ousted by the Dutch as far back as 1642. The Dutch in turn left the Gold Coast at the beginning of the

nineteenth century, and since this was the only part of the region
where they had any posts they thus removed themselves completely
from the West African scene, leaving behind the British, the French,
the Portuguese and the Spaniards to be joined later by the Germans,
who made their first serious acquisitions about the middle of the
nineteenth century.

The most dramatic turn of events in the political evolution of West
Africa occurred towards the end of the nineteenth century as a result of
the Berlin Conference held in 1884–85 at the instance of Germany and
France. The slave trade had been abolished, thanks largely to the
initiative of Great Britain, and West Africa had begun to assume a new
importance as a source of industrial raw materials, especially palm oil
and minerals, for the factories of Europe. For this new trade the mere
control of trading posts was not enough; it was necessary for the
European powers to acquire actual territorial possessions, and the
purpose of the Berlin Conference was to lay down ground rules for the
partition of the continent. Before the Conference the acquisition of
colonies had been a fairly leisurely process, but by enunciating the
principle that in future no claim to territory in Africa by any European
power would be recognised unless that power could show that it was in
effective control of the territory concerned the Berlin Conference paved
the way for the 'scramble' for the continent. Within twenty years the
political map of Africa which had been empty as far as Europe was
concerned became a crazy patchwork of colonial possessions.

The actual course which the scramble took was a rather complex one
as far as West Africa was concerned, events in the region being intimately linked with political events in Europe, which during that period
was itself undergoing rapid political change. By the end of the
nineteenth century, however, the British had established themselves
fairly securely in The Gambia, the Gold Coast, Nigeria and in Sierra
Leone, whose coastal section had been acquired as far back as 1787 as a
colony for the settlement of Negro slaves who had been freed by Britain
or had gained their freedom by escaping to Britain from America; while
the French held the island of Goree and the coastal stretch between the
mouth of the Senegal and the Cape Verde peninsula together with the
extensive hinterland comprising the open and disease-free Sahel and
Sudan zones of West Africa with corridors extending southwards into
Guinea, the Ivory Coast and Dahomey. Germany, which was the latest
arrival in the region, held Togo and the Cameroons, while Spain and
Portugal retained their hold of the group of small islands in the Bight of
Biafra, Portuguese Guinea and Spanish Sahara, consisting of the desert
strip along the Atlantic between Cape Blanco and Morocco. Finally,
there was Liberia, located between Sierra Leone and the Ivory Coast,
which had been founded initially as a small settlement in 1820 by the

American Colonisation Society for the purpose of settling freed Negro slaves from America and which became a fully independent republic in 1847.

Although Britain was at the time the most powerful of the European powers in West Africa as well as in the rest of the continent, it was the French who emerged after the partition as the power with the most extensive territory in the region. The reason was that unlike the other powers they set out deliberately to establish a large African empire mainly for reasons of prestige, especially in view of the setbacks they had suffered in Europe following their defeat in the Franco-Prussian war of 1870–71. The original design of the French, which failed to materialise in the event, was to extend their West African empire right across the continent to their Red Sea base at Obock, in what eventually became French Somaliland. It appears that because of the preoccupation of the British with their Guinea coast possessions, which were economically far more valuable than the interior region sought by the French, the latter were able to have their way without much interference and were thus able in a relatively short space of time by means of a series of brilliant though often ruthless military campaigns to consolidate their position in the West African interior before either the British or the Germans were able to penetrate the difficult forest zone along the coast and drive northwards into the more open and easier terrain of the savanna and Sahel hinterland.

Thus, by the beginning of the First World War in 1914 the outlines of the present political map of West Africa had been laid down and all that remained was for the colonial powers to consolidate their respective possessions. The only significant change occurred after the war, when as a result of her defeat Germany lost Togoland, which was partitioned into two and transferred to Britain and France as Mandated Territories under the League of Nations.

The partition of West Africa into distinct territorial units ruled by different colonial powers had the effect of bringing several different tribal groups together for the first time under a single external political authority. But since the new political boundaries had been drawn without any serious reference to the wishes or interests of the local populations the resulting political units turned out to be no more than mere geographical expressions totally unrelated to the ethnic and socio-economic realities of the region.

The many serious inadequacies of the boundaries drawn by the European powers not only in West Africa but also in the rest of the continent have been attributed both to the excessive preoccupation of these powers at the time of the partition with their own political and economic interests and the haste with which they undertook the two vital boundary-making processes of demarcation and delimitation. In order to complete these two processes in the shortest possible time the colonial powers resorted to the use of the river and stream courses, as well as lines of latitude and longitude and arcs of circles in marking off the boundaries of their possessions, thus sowing the seeds for most of the subsequent conflicts and problems created by boundaries in the region.

But we must be careful not to exaggerate the defects of these boundaries or the harmful social and economic effects caused by them by considering them out of their proper geographical and historical context. Quite obviously in the more thinly populated or desert parts of West Africa the drawing of mathematical or geometrical boundaries could not have caused much serious social damage. Also, although the use of rivers as boundaries is now generally frowned upon on the grounds that rivers and their basins tend to have a unifying rather than a divisive effect upon riparian communities, there is no doubt that many of the large or even moderately sized rivers in West Africa and other parts of the continent did at the time and even today still constitute serious barriers to movement and were thus often regarded by the local inhabitants themselves as convenient features for the demarcation of their boundaries. Indeed, subsequent events have shown that the most serious disruption caused by the colonial boundaries arose not so much from the physical defects of the boundaries themselves as from the fact that in some cases they forcibly brought together within the same political units and cultural moulds imposed from outside people who in most instances had never felt any special affinity with one another or any sense of belonging to the same social and political framework, while in other cases they split up peoples who were culturally homogeneous.

Convenient though rivers are for the demarcation of boundaries, their use for the purpose by the European powers has created a number of problems for several African countries by depriving them of the exclusive control and use of important rivers and their basins. Good examples are the Senegal, which flows between Mauritania and Senegal, the Volta, which flows between Ghana, Upper Volta and the Ivory Coast, and the Gambia, which traverses Senegal and The Gambia. In such a situation no development of importance can be embarked upon unless there is full agreement and cooperation between the states bordering the river, and any joint ventures, such as hydro-electric power projects, that take place can easily be jeopardised if there is disagreement between the states concerned. On the other hand, it is possible to argue that the sharing of a river by two states along their common boundary is a good thing in the long run inasmuch as it can compel them to cooperate with each other in certain vital economic areas and to avoid needless conflict.

The boundaries of West Africa during the colonial period were of three main types: firstly, the international boundaries which marked the limits of the various colonial territories; secondly, the administrative boundaries drawn in accordance with the convenience of the moment to delimit internal administrative divisions, such as provinces and districts; and, thirdly, the intra-territorial boundaries which the French employed for delimiting the constituent colonies of their West African empire which was administered as a single federation with its capital at Dakar. These intra-territorial boundaries, notably that between the Ivory Coast and Upper Volta, were adjusted from time to time as administrative needs dictated, but assumed the stature of international boundaries with the break-up of the federation following the granting of independence.

Despite their many imperfections the international boundaries of West Africa persisted without alteration throughout the period of European colonial rule, apart from a few changes which were either made or attempted in the interest of geographical and administrative tidiness. For example in 1876 negotiations took place between Britain and France for the handing over by Britain to France of The Gambia in exchange for 'any positions held by France between the Pongas river to the north of Sierra Leone and the Equator', but the proposal fell through largely because certain influential circles in Britain were opposed to the idea of giving up the Gambia River long held by the English and of surrendering to a Catholic power large numbers of Gambian Wesleyans (Lucas 1913).

PROBLEMS AND CHALLENGES OF COLONIAL RULE

The period between the Berlin Conference and the end of the First World War was essentially one of boundary making and territorial consolidation in West Africa. By 1922, after the League of Nations conferred mandates upon France and Britain in respect of the former German possessions in Togo and the Cameroons, the outline of the present political map of West Africa had been laid down and the colonial powers were beginning to turn their attention to the internal administration of their territories.

The powers that faced the most formidable tasks in this respect were Britain and France which controlled the greater portion of the land area as well as the population of West Africa. For the lesser powers, namely, Portugal and Spain, things continued very much as they had done in the past, while Liberia found itself in the very special position of an independent black state, free to manage its internal affairs and develop its own institutions without the impediments but also without any of the benefits of colonial rule, apart from American commercial interests in the country's rubber industry owned by the Firestone Company which held large concessions.

The problems which France faced in the administration of her West African empire arose principally from the large size of the empire and the fact that the greater part of it lay in the poor interior regions of West Africa which were far removed from the active economic centres along the coast.

Altogether French West Africa covered an area of some 4.7 million sq. km (1.8 million sq. miles), that is, slightly over 73 per cent of the entire area of West Africa, and eight times the size of France, and in its widest portion measured approximately 3,622 km (2,250 miles) from east to west and 3,059 km (1,900 miles) from north to south (Fig. 12). Distance, together with its strategic and logistical implications, was thus a perpetual problem. And even though the French made determined efforts to overcome it by establishing an extensive network of communications based on railways, roads and waterways, it remained one of the empire's most serious handicaps.

Another problem was the federal form of government by which France sought to administer her empire. Though theoretically sound, this mode of government presented France with the virtually impossible task of holding together within a single political system a wide diversity of peoples drawn from a multiplicity of ethnic origins and geographical backgrounds, all of whom were governed from Dakar, the federal capital, where the governor-general of the federation resided. The result was that even though the federation was divided into a number of colonies each with its own governor, economic and social development tended to be very uneven in its distribution and most attention was concentrated upon Dakar and the ancient colony of Senegal. In the event the only parts of the federation, apart from Dakar and Senegal, which really felt the benefits of France's oft-proclaimed 'civilising mission' were the few large urban centres located in the more prosperous coastal areas, such as Conakry, Abidjan, Lomé, Cotonou and Porto Novo. Thus, although the Federation appeared in name to form a single, all-embracing political entity, in fact it had very little internal cohesion and certainly no spontaneous feeling of unity among its diverse human elements.

Even at the local level much of the traditional cohesion which the region possessed was unwittingly destroyed by the French through the arbitrary manner in which they reduced the power and influence of the local chiefs, who had hitherto exercised undivided authority in religious, economic and political matters within the areas under their control. After a brief attempt at setting up protectorates based on ethnic groupings, they abandoned the idea and resorted to direct administration based on arbitrarily drawn territorial divisions. The

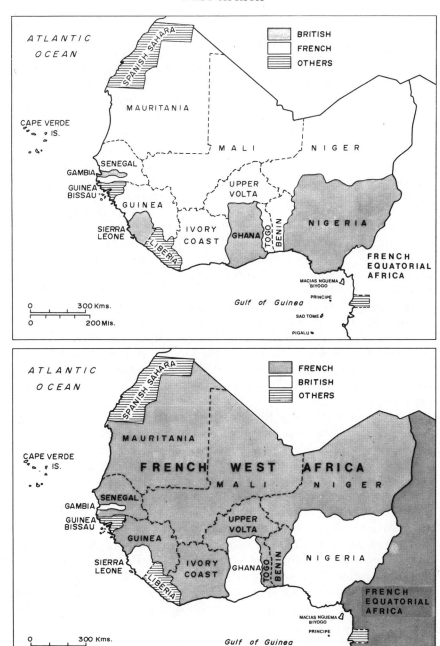

Fig. 12. Contrasted impressions of British and French West Africa

paramount chiefs were systematically removed from office, ostensibly in order to liberate their hapless subjects from their tyrannical rule, leaving only the lesser chiefs, who were shorn of all their traditional powers and functions, except those relating to religion, and reduced to the status of mere functionaries of the government, discharging such mundane but often invidious responsibilities as the collection of taxes, the recruitment of labourers and soldiers for the government, and the settling of minor local disputes. The French, however, followed a different policy in regard to the small group of intellectuals in their colonies, who were accorded special privileges and exempted wholly or partially from several of the fiscal, military and labour obligations imposed on the rest of the population. Unfortunately, the privileged intellectuals or 'évolués' did not do much to help their less fortunate compatriots and remained throughout as a class apart.

In contrast, the problems which Britain faced in her West African empire were relatively simple. British West Africa covered some 1.2 million sq. km (480,000 sq. miles) and consisted of four colonies, The Gambia, Sierra Leone, the Gold Coast and Nigeria, strung along the entire length of the Guinea coast between Cape Verde and the Bight of Biafra. However, each colony was administered as a separate entity under its own governor who was directly responsible to the British government in London. Moreover, the indirect system of government which the British adopted enabled the local rulers and traditional institutions within each colony to be employed effectively as ancillary arms of government.

Like the French, the British also established systems of communications within each colony, although the distances involved were much more manageable and the railways and roads that were built were much more consciously geared to economic development than to purely strategic objectives. In almost every case the overriding consideration was to make each colony as economically self-sufficient as possible without any attempt at internal rationalisation within the West African empire as a whole.

The net effect was that development was far more evenly spread among the four colonies than was the case in French West Africa, where the rationale of government was to develop the federation as an overseas extension of metropolitan France rather than as a group of individual colonies each capable of evolving one day into a fully independent state standing on its own feet.

In both French and British West Africa one of the most serious problems which faced the colonial governments was the difficulty of establishing an adequate infrastructure for effective development, including the provision of social, health and educational facilities. Since productivity was generally low and the amount of money available for investment was very limited, the pace of development remained accordingly slow except in a few specially favoured areas, such as the Gold Coast, which were well endowed with readily exploitable mineral and agricultural resources. The least-favoured areas were the interior savanna, especially the Sudan savanna, where the lack of water and remoteness from the sea proved a serious handicap even to agricultural development for which they seemed to have several obvious natural advantages, especially in the sphere of livestock breeding (Fig. 13).

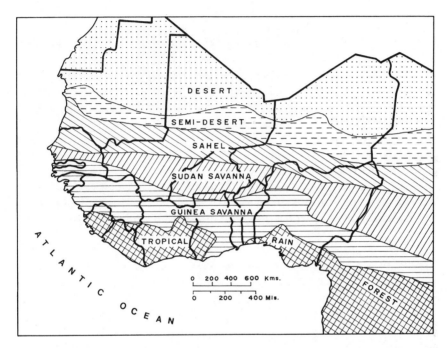

Fig. 13. West Africa: the 'pattern of Nature' versus the 'pattern of Culture' (After Derwent Whittlesey)

But even in the more favoured coastal areas, particularly the four British colonies of The Gambia, Sierra Leone, the Gold Coast and Nigeria, the tendency was to concentrate on the production of only a few minerals and agricultural crops for which the colonies concerned appeared to have clear comparative advantages in terms of the overall world-wide economic pattern of the imperial system to which they happened to belong. Thus, while the economic development of the colonies made sense from the point of view of the controlling metropolitan powers, it resulted in exposing the individual colonies unduly and often dangerously to the vagaries of world commodity markets.

While colonial rule lasted, therefore, the economic role assigned to practically all the West African countries without exception was the production of raw materials in the form of minerals and agricultural crops in exchange for manufactured goods imported from the metropolitan countries.

Despite all these problems and administrative shortcomings, the British and French succeeded eminently in establishing a commendable degree of law and order in their respective empires as a basis for future development, and like the Romans in their Mediterranean and adjoining empires in Europe, Africa and Asia Minor, they could also boast of a 'pax Britannica' and a 'pax Franca' in West Africa, capable of providing in varying degrees a basis for future nation building. And even though the boundaries they drew up in the region were ill-adjusted in most cases to important socio-economic realities and often in conflict with what Derwent Whittlesey described as the 'pattern of nature' (Whittlesey 1938), the fact that these two major powers were on reasonable political terms with each other served to lend stability to the boundaries of their respective colonial possessions in West Africa (Fig. 14).

We have already referred to the importance of location in relation to the sea coast during the colonial period and the reversal of West Africa's trading outlook from its former northward orientation to a southward seaward orientation which took place following the arrival

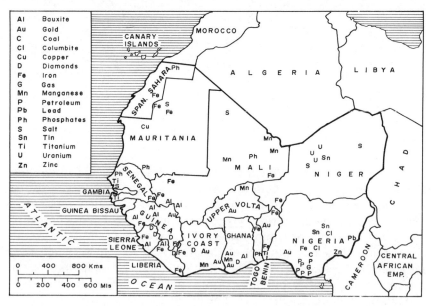

Fig. 14. West Africa: distribution of major mineral resources

of Europeans along the Guinea coast. This change had the effect not only of concentrating trade but also political and social development along the coast. All the initial colonies established by the British and the French were on the coast and served as springboards for further advance inland after the Berlin Conference. The emerging political pattern came therefore to consist of well-established colonies around the various strongholds or forts along the coast, followed inland by protectorates acquired through treaties and other agreements with the local rulers.

Although most of these protectorates gradually lost their special status and were administered more or less as integral parts of the original colonies, they tended throughout the period of colonial rule to lag behind in economic and social development and thus to view their future in a different light from that in which the more highly developed and politically sophisticated coastal peoples saw their future.

EMERGENCE OF INDEPENDENT STATES

The independence movement in West Africa began shortly after the end of the Second World War in 1946. Up till then the only independent country in the region was Liberia which had enjoyed full political autonomy right from its foundation in 1847. There were many reasons for the political changes which followed the war. In the first place, Britain and France, the two principal colonial powers in West Africa, had received very considerable support during the war from their West African colonies in the form of both men and essential raw materials, and many African soldiers had fought on the side of the Allies against Japan in the hot, humid jungles of South-East Asia. In return for these services and also in keeping with the spirit of the times, the colonial powers openly declared their intention to give their colonies greater freedom and better economic opportunities once victory had been won. It was not surprising, therefore, that as soon as the war ended in 1946 and the African troops began to return home from their overseas service with new experiences and increased self-confidence a new wave of political self-consciousness began to sweep across the continent, especially West Africa, reinforced, no doubt, by Britain's imminent granting of self-government to India, Pakistan, Burma and Ceylon.

The first country to achieve self-government in the region was the British colony of the Gold Coast, which became a fully independent state on 6 March 1957 under the new name, Ghana. A relatively small and prosperous country without too many internal divisions, Ghana had in the course of a little over one hundred years of British rule developed a reasonable infrastructure for economic and political advancement and a sizeable number of highly educated and articulate professional men who were united and insistent in their demand for political freedom. The British government conceded this demand without much opposition. Historically, this was an event of great significance in that for the first time a Black African country had successfully won its independence from a major colonial power; and this set a pattern for other colonies to follow and gave them increased moral strength for their fight for self-government. By the end of 1960 all the colonies of the former French West African empire, following Ghana's example, had also achieved their independence as separate republics, and Nigeria, too, had become independent. Then in 1961 and 1965, respectively, Sierra Leone and The Gambia also became independent.

That the pace of change was much faster in French West Africa than in British West Africa despite the fact that the British colonies were generally more developed both economically and politically was due to two main reasons. Firstly, the French had always regarded their West African empire not as a completely separate political entity but merely as an overseas extension of France and they consequently expected, even when African nationalism was at its height, that the colonies would opt for membership of the French Union as proposed by General de Gaulle in 1958 rather than outright independence as separate African republics. Guinea was the first colony to reject the offer and was immediately cut off by France and granted independence; but within two years the other colonies after a brief initial period of vacillation also followed the same course but without breaking all their former ties with France as had happened in the case of Guinea. In other words, instead of allowing her West African colonies to fight individually for independence the French virtually made an offer of independence to them. The second reason, which is closely related to the first, was that whereas the British insisted on a more deliberate process of preparation for independence involving a number of preliminary stages before the final transfer of power, the French made no such demands and thus the transition from colonial status to full political sovereignty was for their colonies a simple and straightforward issue.

Although independence was achieved in both French and British West Africa with such apparent ease, it was preceded in almost every case by a considerable amount of internal dissension often based on tribal cleavage lines, but sometimes also, especially in the case of the former British territories, on the age-old differences and animosities between the inhabitants of the more highly developed and politically sophisticated coastal areas which were the original colonies and those of the less developed interior areas who were more conservative in their outlook and generally suspicious of their coastal neighbours. Nowhere was this contrast more marked than in Sierra Leone, where the Creoles who occupied the colony were descended from repatriated slaves from America and England and were thus quite different in origin and culture from the 'natives', as the Creoles referred to them, who inhabited the protectorate, which not only covered a much larger area and contained the bulk of the country's population but also contained most of its economic resources. Not until proper safeguards for the people of the protectorate could be agreed upon did the British hand over power to the people of Sierra Leone; and in varying degrees similar precautions were taken in all the other British colonies.

For the Portuguese and Spanish territories things remained more or less as they had been for the previous five centuries and the post-war independence movement completely passed them by. However, in 1963 the peoples of Portuguese Guinea began an armed rebellion aimed at securing political independence for themselves. It took ten years for their sacrifices to bear concrete fruit; on 24 September 1973, Portuguese Guinea (now known as Guinea Bissau to distinguish it from the former French Guinea) declared itself independent and was immediately recognised by practically all the African countries and many other member states of the United Nations. On 2 September 1974, Portugal, which had itself in the meantime succeeded in overthrowing the dictatorial government which had ruled the country since Dr Salazar's accession to power in 1932, agreed to hold a meeting with the freedom fighters of Guinea Bissau regarding the granting of independence; and on 10 September 1974 Guinea Bissau was formally declared an independent republic. However, the question of the future of the Cape Verde Islands lying only a few miles off the African coast, which had joined Guinea Bissau in the nationalist struggle and had thus become closely identified with it, was not decided until some months later. On 5 July 1975, Cape Verde was granted independence by Portugal; but it is expected that the question of federation with Guinea Bissau, which had been one of the expressed aims of the nationalist movement, will be taken up at a later date. In the meantime the people of Cape Verde have a great deal to do to put the economy of their small country on a reasonably sound footing after centuries of neglect by Portugal. With Portugal's withdrawal from West Africa, the only part of the region that was still under European colonial rule up to the beginning of 1976 was Spanish Sahara; but even here there were clear signs of impending change.

Although Spanish rule over this territory dates back to 1509, in fact very little development took place and the outside world was kept in almost complete ignorance of its internal affairs. Nor was it clear just what Spain stood to gain from its control, since the 102,000 or so square miles of land comprised within its boundaries were wholly desert and it did not appear to be endowed with any significant mineral resources, while its meagre population, numbering 76,425 in 1970, consisted mostly of impoverished nomads. Following the discovery of rich oil deposits in the adjoining Saharan parts of French North Africa in the middle 1950s the Spanish embarked on the large-scale exploration of oil in the territory but were obliged to curtail operations on account of poor initial results and local opposition from certain nationalist elements. Later, however, the territory's economic prospects took a new turn with the discovery of rich phosphate deposits in the northern portion adjoining Morocco. This led Morocco, which had for years claimed both Spanish Sahara and Mauritania as part of its legitimate political sphere, to renew its demand for the annexation of the territory.

In furtherance of this claim Morocco, towards the end of 1975, when Spain was preoccupied with serious domestic problems arising from the choice of a successor to General Franco, organised a march of Moroccan citizens into Spanish Sahara and eventually got Spain to agree to the transfer of the territory to Morocco and Mauritania without any reference to the local inhabitants. This led to a political clash between Morocco and Algeria, which borders the territory along the north-east and also had territorial ambitions in the area. In the midst of this explosive situation, Spain decided towards the end of February 1976 to terminate its political responsibility for Spanish Sahara, leaving the three principal contestants, namely, Morocco, Mauritania and Algeria, to fight it out among themselves.

So far neither the Organisation of African Unity nor the United Nations has succeeded in getting the three countries involved to agree to an amicable settlement. In the meantime the local nationalist movement of Spanish Sahara, known as the Polisario Front, which enjoys the backing of Algeria, has declared the Spanish Sahara an independent state and is making desperate efforts to secure recognition from the Organisation of African Unity, some of whose members have already given the new state their official recognition.

The developments in Spanish Sahara are of particular importance because they represent the first instance in post-colonial Africa of African countries forcibly annexing an African colonial territory with the connivance of a European power and thus preventing the people of the territory from exercising their natural right of political self-determination and achieving their independence from their former colonial rulers. How the problem will be resolved still remains to be seen; Spanish rule in the territory has admittedly been ended, but the indigenous inhabitants have so far been cheated in the most blatant way of their natural right to independence and political sovereignty and their territory has been partitioned without any reference to them by their two independent neighbours, Morocco and Mauritania.

The problem of Spanish Sahara or, as it is now more usually referred to, Western Sahara, is a very serious one and constitutes a potential threat to peace and stability in the Maghreb and the north-western part of Africa generally. Already it has led to a straining of relations between Algeria, which supports the Polisario Front in its bid for independence, and Morocco and Mauritania, whose occupation of the country is regarded as illegal. To press its claims the Polisario Front has embarked on guerrilla activities which promise to grow in scale unless an early solution is found to the problem. One can only hope that the Organisation of African Unity will rise to the occasion and give the matter the urgent attention it deserves on the basis of justice and fair play and in keeping with the spirit of the Organisation's charter so that the people of Western Sahara, like the citizens of other independent African countries, can experience for themselves the true meaning of political emancipation and determine their own political destiny.

It is noteworthy that, as in other parts of Africa, it is within the framework of the boundaries and territories established by the colonial powers during the partition of the continent that the newly independent states of West Africa have been obliged to express their new political identities. It seems clear that despite all their imperfections these colonial boundaries are likely to persist for a very long time, and such changes as are made will consist mainly of minor local adjustments aimed at settling incidental differences between neighbouring states, especially where the course of the boundary affects the ownership of important economic resources such as oil and other minerals or grazing rights in areas where pastoralism is a major economic activity.

In general, therefore, it can be said that the effect of the European colonisation of West Africa – and indeed of the rest of the continent south of the Sahara – has been to bring into being a series of young nation states that are faced with the very difficult task of adjusting themselves to the artificial boundaries imposed by the colonial powers in the process of partitioning the continent among themselves.

CONTEMPORARY PROBLEMS OF NATION BUILDING

One of the most obvious defects of the new states of West Africa is the artificial nature of most of their boundaries. Yet surprisingly, although a few boundary disputes have arisen among these states, such as the current one between Mali and Upper Volta, considering the nature and origin of most of the boundaries within the region, the number of disputes has been remarkably small. Partly this is due to the stabilising influence of the Organisation of African Unity founded in 1963 which through its mediation and arbitration commission has intervened successfully in several disputes not only in West Africa but in other parts of the continent and partly to the fact that the problem of arbitrary and ill-adjusted boundaries is so widespread in the region that most of the newly independent states have recognised the futility of attempting their rectification and have decided instead to accept them as necessary evils with which they must learn to live.

Fortunately, also, such questions as the possibility of external attack and large-scale international conflicts have never been serious issues in the region. The issues which have presented really serious problems to the new states have been such questions as internal cohesion and unity, economic viability and political stability.

It was hardly to be expected that the internal cohesion of the various ethnic groups or tribes within their territories and the sense of national unity felt by their subjects would be among the principal preoccupations of the colonial powers. Indeed, from their point of view a certain amount of internal division was useful as a means of reducing or neutralising local opposition to their rule. Accordingly, the task of welding their various peoples together and of giving them that sense of unity which is essential for effective nationhood has had to fall to the newly independent states.

We have already referred to the age-old conflicts between many of the coastal peoples and those of the interior in most of the colonial territories. Since independence tribal animosities based on petty rivalries and jealousies often created by actual or imagined injustices in the sharing of various benefits, especially the spoils of political office, have assumed serious proportions in several countries. The most outstanding example is Nigeria, where the jealousies between the Yoruba and Ibo of the south and the Hausa of the Moslem north led to serious bloodshed in 1966 and eventually developed into a bitter civil war lasting from 1967 to 1970 between the Ibo, who threatened secession, and the rest of the country.

On the eve of the civil war the Nigerian government decided to split up the country's three regions, upon which the federal constitution had been based at the time of independence, into twelve new states more comparable in size and population to one another. This was intended to put an end to the gross imbalance in size and population that had existed among the regions, whose boundaries largely coincided with the three principal ethnic groupings, and also to curtail the very considerable political and economic power previously enjoyed by the three regions in relation to the federal government itself. It was felt that these defects in the country's regional set-up had been the main cause of the political instability and internal unrest that had marked the post-independence period and that the creation of the twelve states would provide the necessary remedy (Figs. 15 and 16). But even this measure, drastic as it was, failed to bring under control the divisive effects of tribalism and produce the expected degree of internal political equilibrium; for some of the new states, especially those in the north, were still disproportionately large. Early in 1976 it was decided as a corrective measure to increase the number of states to nineteen and move the national capital from Lagos to a more central and politically neutral location in order to make it more easily accessible to all parts of the country (Fig. 17). However, before these changes could be implemented the head of the government responsible for the proposals, which only seven months earlier had ousted the previous government through a military coup, was assassinated in an abortive military coup. Although, as was expected, his successor pressed the changes through, it is highly doubtful if they will be sufficient in themselves to bring under effective control, let alone eradicate, the tribal jealousies and animosities which appear to have become such a deeply rooted feature of Nigerian political life.

Despite several efforts at playing down its importance, tribalism remains a serious divisive force in most of the new states of West Africa, and really bold and imaginative measures will need to be taken in order to eliminate it completely or, at least, to diminish its effects. Among the principal reasons for its persistence is the fact that it is so intimately

Fig. 15. Nigeria: the three regions at independence

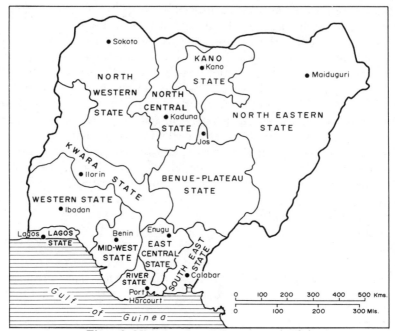

Fig. 16. Nigeria: the twelve states (1967)

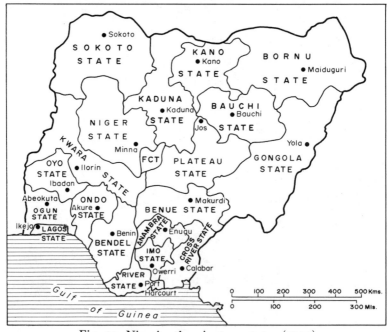

Fig. 17. Nigeria: the nineteen states (1975)

bound up with many of the indigenous institutions and cultural traditions, such as chieftaincy and local languages, which are now being deliberately stressed in many countries as part of the drive to develop national consciousness and pride based on distinctively African foundations. Until such obvious contradictions have been resolved it will not be possible to identify and eliminate the negative aspects of tribalism and nurture its more positive and constructive aspects for the difficult task of national integration.

The problem of economic viability is one that has plagued most countries that have recently emerged from colonial rule. The point at issue is not mainly or simply that their economic output and *per capita* incomes are low by comparison with those of the developed countries, but that the structure of their production and trade is grossly unbalanced and based on the export of a few raw materials of mineral or agricultural origin in exchange for manufactured goods supplied by the industrialised nations. This was the pattern during colonial times and it has persisted with very little change following the attainment of independence, thus exposing many of the new states of the region unduly to the vagaries of the world commodity markets. One of the things which political independence was expected to offer was increased material prosperity not in order to enable the mass of the people to lead a more luxurious life but in order to make it possible for them to enjoy the bare necessities of life. The failure of most of the indigenous governments of West Africa to fulfil these expectations has led in many cases to serious disillusionment with the whole idea of self-government.

In the face of current world economic trends and the rapid technological advances of the more highly developed and richer nations the task of economic diversification and improvement with a view to improving the living standards of their people has become even more difficult for the new African states; and their problems have been further compounded in recent years by the inflationary pressures generated by the world oil crisis. Thus, without the assistance of unexpected bonanzas like rich oil finds or other valuable mineral discoveries very few of the new states of West Africa or other parts of the continent are likely to develop particularly strong economic foundations. There is still, nevertheless, the possibility of improving considerably upon the fragile and unbalanced economies bequeathed to them by the colonial powers, given hard work and realistic planning.

TERRITORIAL AND STRATEGIC ASPECTS

From a global point of view West Africa does not and has never occupied a position of any great strategic importance in respect of its geographical location, although in a few instances in the course of its history it has played a peripheral role in the conduct of military operations within and around the North Atlantic. Nevertheless, it is a region of considerable strategic importance in terms of its economic resources, which include a number of strategically vital raw materials, such as vegetable oils, rubber, high-quality iron ore, manganese, copper, bauxite, tin, columbite and, above all today, petroleum (Fig. 14, p. 116).

Before the European discovery of the Guinea coast in the fifteenth century the region formed a relatively isolated corner of the African continent, cut off from the rest of the world by the Atlantic Ocean along the west and south and virtually barred from the rest of Africa itself by the Sahara on the north and the difficult mountain and forest terrain of equatorial Africa along its eastern border. It was the development of the slave trade and the trade in gold and spices that gave it its first real links with Europe and America; but after the abolition of the slave trade the American connection virtually disappeared and the region's external contacts became almost wholly confined to trade with Europe.

After the partition of the continent, however, the main attention of the British and the French, who held the greater portion of West Africa, became focused upon its internal problems, especially those associated with the control and administration of its vast area, the penetration of its generally inaccessible hinterland and the effective conquest of its vast interior distances from a seaboard notoriously poor in natural harbours and inlets.

The problems which faced these two powers were not entirely the same; for while the French West African empire formed a continuous area which could be served by a single and continuous system of land communications supplemented by navigable waterways, the only effective means open to the British for establishing direct contact between their widely scattered colonies was by ocean transport. Thus, throughout the period of colonial rule French West Africa remained essentially a continental or land-based empire, while British West Africa, with the possible exception of Nigeria which both covered a very large area and extended far into the interior, was essentially coastal in nature (Fig. 12, p. 113). Owing to the physical limitations of the Guinea coast, however, few really outstanding harbours were developed by either the British or the French, with the exception of Dakar, which the French developed into a port and naval base comparable in importance to some of the leading ports in France itself, and Freetown, whose natural advantages were such that it could be used as a naval base by the British in times of emergency, even though as a port its facilities, unlike those of Dakar, were rather limited.

The Second World War gave West Africa a new geo-strategic significance, following the Axis occupation of Southern Europe and

North Africa and the blockade of the Mediterranean and Suez route linking Western Europe with the Middle East, India, South-East Asia and Australia. Thanks to the new developments in air transport, the region became an important staging post for the ferrying of vital war supplies from America to the North African theatre as well as to the eastern part of the continent. In terms of sea transport, too, Dakar, Freetown and some of the other ports along the Guinea coast, such as Takoradi and Apapa, became important stopping points on the alternative route from Europe and America to the Indian Ocean round the Cape of Good Hope.

Before the war air transport had been negligible in West Africa; but as a result of wartime exigencies several airports such as Yundum at Bathurst, Hastings, Wellington and Lungi, near Freetown, Takoradi and Accra in the Gold Coast, and Apapa, Kano and Maiduguri in Nigeria, were either developed from scratch or vastly improved. Also, following the breakaway of French West Africa from the Vichy regime then in control of France in 1943, the fine Dakar airport at Yoff was added to the Allied chain of strategic airports in West Africa (Harrison Church 1961).

Apart from its purely logistical role, West Africa, especially the British colonies, served during the war as an important source of vital food supplies and essential raw materials for the war effort, such as chromium, iron ore and palm kernels from Sierra Leone, manganese, bauxite, palm kernels and cocoa from the Gold Coast, and tin, columbite, rubber, palm kernels and groundnuts from Nigeria.

After the war the region reverted to its traditional economic and political role in relation to Europe and the rest of the world, although it was obvious that profound changes, especially in the political sphere, were imminent. But in the field of air transport some of the wartime airfields, such as those at Dakar, Freetown, Accra, Takoradi, Lagos and Kano, remained as permanent legacies and provided an invaluable foundation for the developments in civil air transport that have since taken place in the region. The war also demonstrated much more clearly than ever before the importance of West Africa in relation to the Cape route between Europe and the Indian Ocean, including the Persian Gulf, especially during times of crisis in the Middle East, when the shorter Suez route is not available for normal use by international shipping lines.

But far the most outstanding development that has taken place in the strategic value of West Africa since the war has been the discovery of rich and extensive oil deposits in south-eastern Nigeria, which ranks today as the fourth largest exporter of oil in the world after Saudi Arabia, Iran and Venezuela, with an annual output estimated in 1974 at $9,200 million and reserves estimated at 15,000 million barrels.

Apart from oil several important mineral discoveries have been made in the region since the war, including some of very great strategic and economic value, such as iron ore in Liberia, Guinea and Mauritania, bauxite in Guinea and Ghana; copper in Mauritania; and phosphates in Togo and Spanish Sahara; and there are indications that as geological exploration advances many more discoveries will come to light (Fig. 14, p. 116). Another area where important developments have occurred in recent years is in the field of hydro-electric power. Before the war no major hydro-electric power projects existed in West Africa; today, however, major schemes have been developed in Guinea, the Ivory Coast, Ghana and Nigeria, which have already begun to make dramatic impacts on the economies of the countries concerned.

It has often been said that Africa is the continent of the future. As far as mineral production and the development of hydro-electric power are concerned, this is certainly true of West Africa, where vast potentialities still remain untapped.

POPULATION AND RESOURCES

West Africa's present population of 115 million is growing at a rate of between 2 and 3 per cent per annum, which is among the highest in the world. The average population density in the region as a whole is around 18 per sq. km (47 persons per sq. mile), but since the northern half consists either of desert or of very dry savanna, most of the population is concentrated in the moister and more productive forest zone along the south, where densities range from 31 to as much as 116 persons per sq. km (80–300 persons per sq. mile) (Fig. 34, p. 255). However, there are a number of places in the Sudan savanna zone bordering the Sahel, e.g. in the Kano–Katsina area in Nigeria and around Ouagadougou in Upper Volta, where very high densities going back several decades are found. On the whole, however, it can be said that the southern part of the region is much more densely populated than the northern part, although within each area the reasons for internal contrasts are not always easy to establish since so many different factors – economic, historical, political, strategic and environmental (most notably the incidence of certain serious diseases such as river blindness and sleeping sickness) – have been responsible for determining the present pattern.

This contrast in economic prosperity and population density between the northern and southern sections of the region is a phenomenon of very long standing, but it has been further accentuated over the years by the rapid economic and social advances that have taken place in the coastal regions, especially in the more prosperous states such as Ghana, Nigeria and the Ivory Coast, as well as by the steady process of desiccation and desertification in the Sahelian and savanna zones of the north which has reached disaster proportions during the past few years, causing widespread famine and the loss of human and animal life on an unprecedented scale. The worst-affected areas are Mauritania, Mali, Upper Volta, the northern and eastern sections of Senegal and certain parts of northern Nigeria.

One striking effect of these internal economic differences has been the development of migrant labour movements from the poorer areas like Mali and Upper Volta into the more prosperous neighbouring states like Ghana, the Ivory Coast and Senegal. This has led, especially in the case of Ghana, to the development of large alien populations who have served as a convenient source of cheap labour in the mining and urban centres as well as in the more prosperous agricultural areas where ready cash employment on cocoa farms is available. Other related phenomena are the seasonal migration of labour from Senegal into The Gambia during the groundnut harvesting season and the more evenly spread migration of labour from the northern parts of Nigeria, which lie within the dry savanna zone of West Africa, into southern Ghana and the southern parts of Nigeria itself.

Reliable statistics relating to these movements are not easy to come by, but it is known for certain that in 1960, when Ghana's total population was approximately 7 million, no less than 800,000 aliens, mostly of West African origin and comprising a large proportion of Nigerians as well as nationals of Upper Volta, were resident in the country.

During the colonial period these migrations and the consequent development of large reservoirs of alien labour in the more prosperous countries hardly created any serious problems and, indeed, were generally welcomed as a means of raising productivity in the recipient countries. However, with the emergence of independent states and the consequent sharpening of political and economic boundaries, a complete change of attitude has taken place towards the uncontrolled influx of aliens. In 1970 Ghana gave the first concrete expression in the region to this change of attitude by enforcing its Aliens Act, which led to the expulsion of thousands of aliens from the country, and since then other countries in the region have adopted similar measures aimed at ensuring that the economic interests of their own nationals are not undermined or compromised by the presence of unduly large numbers of resident aliens.

These measures have had far-reaching political repercussions throughout the region and led in several instances to strained relations among the countries most directly affected. But they have had some beneficial effects and have at least succeeded in drawing attention to the need for each country to aim at self-reliance in matters of employment and labour supply. No discussion of the distribution of population in

West Africa can be complete without a special reference to Nigeria, whose population estimated at nearly 63 million in 1975 makes it the most populous country in West Africa (with some 54.6 per cent of the region's total population) as well as in the whole of Africa and the ninth most populous country in the world. This fact, coupled with the country's recently discovered wealth in oil, gives it a political and strategic importance in West Africa which no other country can rival, even though in terms of *per capita* income it is still not the most prosperous country in the region. Indeed Nigeria has the potential to become the most powerful country in Africa in both political and economic terms.

Apart from the problems created by population imbalances and movements between states, practically all the countries of West Africa, especially the more prosperous ones, are today faced in a more acute form than ever before with the problem of rural depopulation and urban drift with all its attendant social and economic consequences within their own boundaries. In many cases urban populations are growing at twice or even thrice the rate of normal population growth in the countries concerned. Between 1960 and 1970, for example, the population of Greater Accra in Ghana rose from 337,000 to 850,000, while that of Lagos, Nigeria's federal capital, rose from about 600,000 to 1 million; and the same tendency is to be found in several of the large towns throughout the region.

The fact that these phenomenal increases in numbers are not matched by a corresponding expansion in employment openings and housing facilities has led to widespread frustration among large sections of the urban population and a general deterioration of the urban environment, all of which pose serious threats to social and political stability in many parts of the region. In many ways these internal threats are far more dangerous than any threats that might be directed from outside, and it is of the utmost importance that prompt and appropriate measures should be taken to tackle their root causes. The mere passing of laws to restrict the emigration of people from the rural areas into urban centres cannot solve the problem; a far more effective solution is to make rural life more attractive and rewarding by ensuring that some of the basic amenities and infrastructural services like hospitals, schools, electricity and good drinking water which are now concentrated in the urban areas are extended to the rural areas through bold and imaginative programmes of rural development.

CLOSER UNION AND INTER-TERRITORIAL COOPERATION

The post-independence period has been marked by several moves for closer union and cooperative arrangements of various kinds both among the countries of West Africa and with other African countries outside the region. But in fact such moves date back at least to 1920 when a handful of intellectuals and political leaders, mostly from the Gold Coast, held the first British West African Conference in Accra. At that time the aim was to awaken African national consciousness and to bring about a federation of the four British colonies as a means of achieving eventual political emancipation from foreign rule, and the French colonies were accordingly not represented since their problems and political attitudes were entirely different.

These early efforts produced little in the way of tangible results, but they succeeded in establishing useful bonds among the African leaders and in drawing public attention to the evils of colonialism. At the official level, however, the mounting tempo of local political agitation and the fact that all the four British colonies were ruled from the Colonial Office in London led to an increased degree of consultation and cooperation among local administrators in several spheres of activity – legal, scientific, technical, economic and political.

During the Second Word War the need for even greater cooperation among the West African colonies became imperative and a special resident minister of cabinet rank with his headquarters in Accra was appointed to coordinate the war effort in the region. After the end of the European war in 1945 the British government decided to continue with the policy of regionalism, and the resident minister was replaced by a West African council composed of the four British governors whose function was to coordinate policy and discuss common problems in the region. By 1946 the regional idea had taken firm root and a number of institutions and bodies were set up to serve the needs of British West Africa in a variety of fields, such as scientific research, air transport and the promotion of industries and the arts (Bourret 1952).

Behind all these developments was the tacit recognition that the time for independence was drawing near and therefore it was necessary to speed up the economic development of the region in keeping with the rapidly changing concept of empire.

Had the British West African colonies formed a continuous stretch of territory like French West Africa or British East Africa, it is very likely that many of these regional bodies might have survived the strong local pressures for the establishment of separate national institutions that built up as a result of the nationalist fervour which followed the achievement of independence. Unfortunately, as soon as the British withdrew from the region all the inter-territorial institutions and arrangements they had established began to crumble and each country proceeded to establish its own national institutions based on its own assessment of national needs and priorities. Nevertheless, their common historical and political experience as former British colonies and their common use of the English language as well as their continued membership of the Commonwealth have so far served to give all of them an identity of interests and purpose in the conduct of their affairs that is perhaps even more important than any formal agreements for closer cooperation that may be concluded among themselves.

In contrast with the former British territories, the French-speaking countries of West Africa have shown a far greater tendency since independence to enter into formal alliances with each other and with other French-speaking African countries outside the region. In the early 1960s, when the African independence movement was at its height, there was an unusual proliferation of such alliances, mainly for political and ideological reasons. Two major groups emerged: the Casablanca Group, consisting of the more radical North African states, namely, Morocco, the United Arab Republic, Algeria, Libya, together with the three most radical West African states, Ghana, Guinea and Mali, and the more moderate and conservative Brazzaville Group, comprising all the French-speaking states of West Africa, with the exception of Guinea and Mali, together with Cameroon, the four countries of former French Equatorial Africa and Madagascar. Within this group was an exclusively West African group known as the Council of Understanding (Conseil d'Entente), which was formed on the initiative of the Ivory Coast a few months before it attained independence in 1960 together with Upper Volta, Dahomey, Niger and, later (1963), Togo with the aim of coordinating their economic and foreign policies and consulting each other regularly on matters of common interest. Later on this Council joined with Senegal and Mauritania to form a common union.

The Brazzaville Group was formed in the autumn of 1959 for the express purpose of discussing the Algerian question, then the most burning issue on the African scene, and of finding ways of mediating between France and the Algerian nationalists. Later on, however, the group became more institutionalised and assumed the name, Union of African States and Madagascar (UAM), which was revised in 1965 as the Afro-Malagasy Common Organisation (OCAM) with slightly modified functions and an extension of its membership to include Congo Leopoldville (now Zaire), which is also a French-speaking country. Among the earliest and most notable economic achievements of this union was the setting up of a joint inter-territorial airline, Air Afrique, which ranks today among Africa's major airlines.

In addition to these two major groupings mention should be made of the Monrovia Group of non-socialist African states which grew out of the Monrovia Conference of twenty African statesmen convened by Liberia in 1961 to serve as a counterpoise to the Casablanca Group. Significantly, the conference was boycotted by Ghana, Guinea and

Mali and the other radical African States outside West Africa. But like the Casablanca Group this group was obliged to discontinue its activities after the establishment of the Organisation of African Unity.

In all the political and ideological groupings that have taken place in West Africa and other parts of the continent since the process of decolonisation began the dominant influence of language and past historical affiliations is quite apparent. This is especially noticeable in the case of the French-speaking countries, which have exhibited a far greater degree of solidarity than their English counterparts. There is no doubt that this is largely due to the highly centralised nature of the former French Union to which they all belonged before independence and the fact that within West Africa all these countries were formerly part of a single federation covering an unbroken stretch of territory. The only union of West African states that has ever succeeded in crossing the English–French linguistic barrier is the short-lived Ghana–Guinea–Mali Union which lasted from 1959 to 1963. But it is significant that it owed its sudden birth no less than its equally sudden demise to the fact that it was created in response to the urges of a fleeting political ideology rather than on the basis of rational economic and political considerations.

Among the English-speaking countries, although good relations have generally prevailed at both official and unofficial levels, the urge for the establishment of formal unions and cooperative arrangements has been much less strong. Indeed, during Nkrumah's rule in Ghana he deliberately opposed all such arrangements on the grounds that they constituted an impediment to the speedy attainment of his pet dream of a union government for the whole of Africa, unless, as in the case of the abortive Ghana–Guinea–Mali union, they were clearly aimed at further extension to cover the entire continent.

Since the heady days of the immediate post-independence period, however, significant changes of attitude have taken place in West Africa towards the question of inter-territorial or intra-regional co-operation. Thanks largely to the influence of the Economic Commission for Africa, the emphasis has shifted from reliance on formal political agreements as a means of achieving economic objectives to a more pragmatic approach based on functional arrangements directed towards specific goals. The most notable example of this new approach so far is the Economic Community of West African States (ECOWAS) formed in 1975 by the independent states of West Africa. No doubt the cultural and political affiliations of the past will continue to exert a strong influence on policy-making for many years to come, but increasing recognition is bound to be given to the importance of economic considerations as a basis for future cooperation among the states of the region.

7

Central Africa

Central Africa is best defined in terms of its political divisions. These consist of ten states or political units which can be grouped broadly into the following three divisions: (a) the former Belgian colony of the Congo and the Trust Territory of Ruanda-Urundi; (b) the former French Equatorial Africa together with the former Trust Territory of French Cameroon and the southern section of its British counterpart, now split up into the independent states of Gabon, the Congo Republic, the Central African Republic, renamed the Central African Empire in December 1976, Chad and the Federal Republic of Cameroon; and (c) the offshore volcanic islands of São Tomé and Principe, recently freed from Portuguese rule, and the islands of Fernando Po and Annobon, now renamed Macias Nguema Biyogo and Pigalu, respectively, together with the mainland enclave of Rio Muni, all formerly under Spanish rule but now forming the independent republic of Equatorial Guinea (Fig. 18). Altogether, as shown in the table below, these divisions cover an area of slightly over 5.4 million sq. km (2 million sq. miles) and contain a population of some 47 million.

Political divisions	Area Sq. km	Sq. miles	Population 1975 or latest
Burundi	27,834	10,747	3,765,000
Cameroon	475,442	183,568	6,398,000
Central African Empire	622,984	240,534	1,720,000
Chad	1,284,000	495,752	4,030,000
Congo	342,000	132,046	1,345,000
Gabon	267,667	103,346	526,000
Equatorial Guinea	28,051	10,831	310,000
Rwanda	26,338	10,169	4,198,000
São Tomé and Principe	964	372	80,000
Zaire	2,345,409	905,562	24,902,472
Total	5,420,689	2,092,927	47,274,472

The region is an unusual one in many respects. Owing to its location in the heart of the African continent and its considerable latitudinal extent from the Tropic of Cancer along the northern border of Chad to Latitude 13°S at the southernmost tip of Zaire only a few miles from the Zambezi River, it shares long land boundaries with all the major

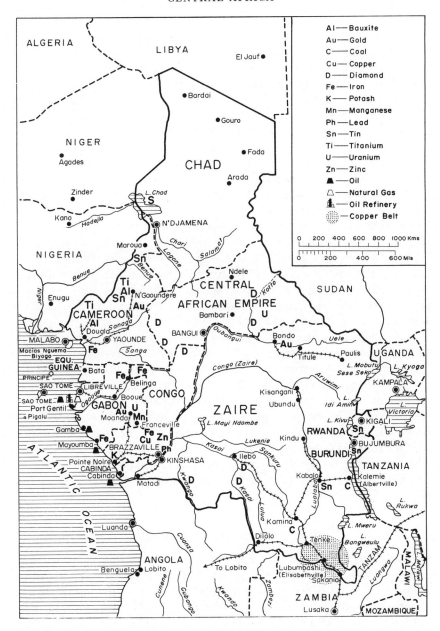

Fig. 18. Central Africa: political divisions, communications and mineral resources

regions of the continent and contains a greater variety of physical environments than can be found in any other region. Practically all the different types of climate and vegetation within the tropical zone of Africa, from the arid wastes of the Sahara desert to the luxuriant forests of the Congo basin with their intervening gradations of steppe, savanna and bush, are represented within its borders, while on the plateau and mountainous rims along its southern, eastern and north-western margins, where altitudes ranging variously between 1,000 and 5,000 m (3,000 and 15,000 feet) occur, the cooler climatic conditions and the vegetation types more commonly associated with Southern and East Africa are also represented.

Indeed, as seen within the total context of the African continent, it is difficult to identify any really valid geographical basis for regarding Central Africa as a distinctive region in its own right. Although individual portions of the region, most notably the Congo basin, may have certain unifying characteristics, the region as a whole is at best no more than an assortment of different fragments detached from neighbouring regions and forming a geographical and political transition zone between West and East Africa, between Mediterranean and Black Africa and between lowland and highland Africa.

In respect of its human geography also, Central Africa exhibits the same degree of heterogeneity that characterises its physical geography. Except along its frontier with West Africa, where Mount Cameroon and the Adamawa Highlands have acted for centuries as a barrier to human movement and racial intermixture between the two regions, Central Africa has enjoyed close human contacts over a long period of time with all the neighbouring regions, especially along its eastern and southern borders. As a result of this, in addition to the Bantu peoples who form over 60 per cent of the present-day population, the region contains a wide diversity of racial and ethnic types, including Arabs, Hausas, Fulani and Nilotic peoples.

MODERN POLITICAL EVOLUTION

As in West Africa, it was the Portuguese and the Spanish who pioneered the modern European colonisation of Central Africa, to be followed later by other European nations. In 1471, the same year in which they made their first appearance in the Gold Coast, the Portuguese discovered the islands of São Tomé and Principe in the Bight of Biafra, although the actual settlement of the two islands did not take place until 1493. Similarly, Fernando Po and Annobon were first discovered and settled by the Spanish in 1472 and 1474, respectively, and developed as bases for the exploitation of tropical products and slaves. Later on, the Spanish also acquired Rio Muni on the mainland, a few miles east of Principe; but their subsequent colonising activities in Africa were cut short by the Treaty of Tordesillas, which gave the Portuguese exclusive rights of occupation in the continent, while conferring similar rights on the Spanish in the New World.

Following the withdrawal of Spain, the Portuguese proceeded to establish a number of coastal possessions on the African mainland in the region of Gabon and at the mouth of the Congo River, which they named the Zaire River, and farther south along the coast of present-day Angola, where they made their first contact with the kingdom of Bakongo, then the most powerful and highly organised of the African nations in this part of the continent. In 1778, however, Spain was allowed by Portugal to resume control of the two islands of Fernando Po and Annobon and to engage in trade along the African coast in the region of Rio Muni in return for certain territorial concessions in the Spanish-held portion of South America.

Portuguese influence in the coastal sections of Central Africa remained paramount until the sixteenth century, when other European rivals in the form of Dutch, English and French merchants appeared. Soon the French succeeded in gaining virtual control of Gabon, whose chiefs gave them full and entire sovereignty over the Gabon River and its adjoining lands between 1842 and 1862, thus placing them in direct competition with the Portuguese in the control of Central Africa (Thompson and Adloff 1960).

All this while the vast interior remained a 'terra incognita' to the European nations, who were quite content to confine themselves to the coastal areas and to rely on African middlemen for the supply of slaves for their settlements in the New World. But the region was by no means empty; it was the scene of intense rivalry among various African kingdoms which were competing with each other for political and territorial supremacy. The most notable of these was the kingdom of the Bakongo, which included the coastal areas around the mouth of the Congo River and the territory immediately to the south, where the Portuguese were busy trying to establish the beginnings of their Angolan empire. As in West Africa, it was not until after the abolition of the slave trade early in the nineteenth century that the European powers seriously began to venture inland with the aim of acquiring large empires which could guarantee for them a share in the natural riches of the interior.

Thanks to the enterprise of explorers like Paul du Chaillu and Savorgnan de Brazza, France was able to establish her position in Gabon and the lands lying north of the Congo River, while King Leopold II of Belgium, through the daring and extensive expeditions of the American explorer, Stanley, who was working as his agent, gained control of practically the whole of the Congo basin, thus forcing the

Portuguese to confine their colonising efforts to the coastal and interior regions south of the Congo estuary.

The final share-out of Central Africa took place at the Berlin Conference of 1884–85, which confirmed France in her possession of the lands lying to the north of the Congo River and its northernmost tributary, the Oubangui, while the Congo basin itself was recognised as the personal empire of King Leopold II.

That King Leopold was able to acquire such a vast empire as his personal property was due to the initial refusal of the Belgian government and the people of Belgium to support him in his bid to secure a share of the African continent for Belgium, which was without any overseas possession. However, recognising the interest of the major powers in the Congo on account of the enormous riches it promised as well as the striking opportunities which its rivers offered for transport in the heart of the continent, King Leopold enlisted the support of these powers in forming the International African Association, later renamed as the Congo International Association, with the ostensible aim of promoting the exploration and civilisation of Central Africa on a purely scientific and philanthropic basis. By assuring the other colonial powers as well as the United States that the Congo basin would be open to free trade and access for all nations, he was able to gain international support for its independent operation, thus confirming his personal ownership of the territory. Later on the Belgian government was persuaded to recognise the king as the head of the new state, which was renamed as the Congo Free State and granted a formal political union with Belgium but without altering its status as the king's personal property.

King Leopold's next task was the territorial consolidation of the Congo. For the most part this proved to be a peaceful though somewhat slow process. The only part where serious opposition was encountered was along the eastern frontier, where a number of Arab chiefs had for a long time established themselves in the slave and ivory trade; but they were successfully defeated by Belgian forces between 1892 and 1894. In the mineral-rich Katanga region in the south-eastern part of the territory, also, special expeditions had to be mounted between 1891 and 1892 to forestall Cecil Rhodes's forces, who were trying to push northwards from Northern Rhodesia.

Unfortunately, however, despite the relatively peaceful manner in which the Congo had been acquired, King Leopold's subsequent administration of the territory quickly degenerated into a cruel and oppressive regime as a result of the ruthless and inhuman methods adopted by his agents in extracting quotas of raw materials, especially rubber, from the local inhabitants and the many atrocities which these agents perpetrated against them. Eventually, under the pressure of outraged international opinion Belgium was obliged in 1908 to take over the territory from the King and administer it as a Belgian colony.

North of the Congo, France's possessions in Central Africa up to the time of the Berlin Conference comprised only Gabon and the territory now forming the Congo Republic. It was not until 1887 that de Brazza claimed the region north of the Oubangui as a zone of French influence in a bid to link up France's possessions in equatorial Africa with French West Africa and Algeria. In furtherance of this scheme the French occupied in 1889 the territory approximately between the Oubangui and the Chari rivers, now forming the Central African Empire, and in 1900 extended their influence northwards to Chad. However, their attempt to advance farther westwards in the region to the south of Lake Chad was halted by the British and the Germans; and in 1890 and 1894, respectively, they were forced to recognise the extension of the eastern boundary of Nigeria and the northern boundary of German Cameroon to the lake. A further attempt was made by the French to extend their possessions eastwards into the Sudan, but they were checked by the British with the aid of a Sudanese adventurer named Rabah (Thompson and Adloff 1960).

Although the French were able to gain possession of Equatorial Africa within a comparatively short period, the actual settlement of the colony's frontier took a considerably longer time and had barely been completed before the First World War broke out in 1914. The war brought about two important changes in Central Africa. As a result of Germany's defeat, the German colony of Cameroons was created a Mandated Territory and partitioned between Britain and France with the northern sixth going to the former and the remaining five-sixths to the latter, while the German East African colony of Tanganyika was also created a Mandated Territory and shared between Britain and Belgium. The area that went to Belgium was the tiny portion in the extreme north-western part known as Ruanda-Urundi, which thus became administratively detached from East Africa and brought within the political framework of Central Africa.

With the removal of Germany from the scene, Central Africa became a predominantly French and Belgian sphere. Although the British, Portuguese and Spanish portions still belonged technically to the region, owing to their location along the north-western borders of the region, their commercial and even cultural affiliations tended to be directed more towards the coastal areas of Eastern Nigeria than towards Central Africa.

PROBLEMS OF ADMINISTRATION AND ECONOMIC DEVELOPMENT

As in West Africa, the main task which faced the two principal powers in Central Africa, namely France and Belgium, following the final settlement of the boundaries of their empires, was the orderly administration and development of their respective territories. French Equatorial Africa and the Belgian Congo had one thing in common: in both of them French was the official language thus facilitating communication between them; but in practically all other respects – cultural as well as physical – the two empires exhibited wide divergencies, apart from the fact that owing to their immense size and ethnic diversity the administration of each of them raised extremely difficult problems, especially in the sphere of transport and communications.

However, the Belgian Congo had a distinct advantage over French Equatorial Africa because of its immense natural wealth and the fact that despite the many interruptions in its course due to falls, cataracts and gorges, the Congo River system offered an excellent basis for the development of an extensive system of communications embracing railways, roads and waterways, which were supplemented after the Second World War by an impressive network of air transport. In contrast, the French, operating in a much more difficult terrain and in an area much less well endowed in terms of natural resources than the Belgian Congo, only succeeded in developing a rudimentary system of transport and communications almost wholly confined to the coastal portions of their empire.

The subsequent economic development of the two empires exhibited similar contrasts. While the French paid relatively scant attention to Equatorial Africa as compared with their larger West African empire, which though not particularly rich in natural resources was far more accessible and strategically important, the Belgians, who had no other colonial possessions outside of Central Africa were able to concentrate their full attention on the Congo and to transform it quickly and methodically into one of tropical Africa's most prosperous and efficiently managed countries.

In their social policies, however, the Belgians were much less progressive than the French. Their attitude towards the Congolese was paternalistic, and although they did a great deal to promote the material welfare of the broad mass of the population by emphasising social services and elementary education, they completely neglected the important areas of higher education and professional training, thus making it virtually impossible for the Congolese to play any serious part in the government of the country or to occupy positions of responsibility in either commerce or industry. Added to all this, they also practised a rigid colour bar, especially in the larger urban centres, where the Africans were confined to rather dingy townships on the city outskirts except during working hours when they were allowed in the European areas.

The French, on the other hand, even though they did not do very much to promote the social advancement of their subjects in Equatorial Africa, generally left the door open for those who had the ability and the desire to advance themselves as far as they wished in practically every walk of life without any restrictions based on race. By comparison with the Belgian Congo, French Equatorial Africa was a poor and economically undeveloped territory, but the inhabitants were much freer than their counterparts in the Belgian Congo and could look forward to a considerably brighter political future than was possible for the Congolese.

These contrasts between the two territories were later to have far-reaching consequences on the course of political and economic events. When the French eventually withdrew from Equatorial Africa the new African governments that succeeded them could at least count on the support of a reasonable cadre of well-trained and experienced Africans in various spheres of activity, while in the case of the Congo independence had to be forcibly wrenched from the Belgians by an African population with hardly any experience in political matters and conspicuously lacking in an educated or professional elite capable of immediately taking over the key positions in government and other areas of public life. On the other hand, the Belgian Congo proved to be far better equipped in material terms for independence than French Equatorial Africa owing to the excellent infrastructure which the Belgians laid down in the form of communications, social services and public utilities and the large amount of capital which they sank into the development of the Congo's extensive natural resources. Thus, once independence had been achieved the Congolese had ready at their disposal an economic and social foundation for nation building which had no comparison in tropical Africa, although the administrative problems involved in its effective utilisation were undoubtedly formidable. What remained was how they would build upon this foundation; and here the record so far has been rather disappointing.

Fortunately for the countries of Central Africa, the question of permanent European settlement was never a serious issue during the period of colonial rule. In French Equatorial Africa and within the vast lowlands of the Congo basin the question was entirely ruled out for climatic reasons. But in the plateau region of Katanga in the southern part of the Belgian Congo as well as in the mountainous region of Ruanda-Urundi and the adjoining eastern lip of the Congo basin, where temperatures were much more moderate, conditions were ideally suited to permanent white settlement, and in fact considerable numbers of Europeans concentrated in these areas. However, Belgian colonial policy consistently discouraged the development of permanent and exclusive European communities such as those which developed in the neighbouring British and Portuguese territories in Highland Africa.

The Belgian residents in the Congo were not granted any special political rights and, like the Congolese themselves, were excluded from any active participation in the government of the territory, which was tightly held in the hands of the Belgian government and its local officials.

However, in both the Belgian Congo and French Equatorial Africa the Belgians and the French freely expropriated land for the purpose of establishing plantations owned by the government or by private European concerns for the production of various crops. Similarly, especially in the Congo, large concessions were granted to state-owned and private companies for mineral exploitation. But despite their excessive preoccupation with economic gain, it must be stated in fairness to the Belgians that they followed very liberal and far-sighted policies in the recruitment of the vast labour forces required for their many commercial enterprises and showed a commendable concern for the material welfare of their African employees, even though the opportunities for advancement which they held out to them were strictly limited to the lower and middle grades of employment.

By comparison with the administrative and economic problems of the Belgian Congo and French Equatorial Africa, the problems that faced the British, the Portuguese and the Spanish in their territories along the northern border of Central Africa were quite simple and straightforward. Although British Cameroons was first a Mandate and later a Trust Territory, it was administered in all respects as an integral part of Nigeria. But its peripheral location along the country's eastern border, far from the main political and economic centres, tended to operate as a handicap to its active development.

In the case of the Portuguese and Spanish islands in the Bight of Biafra and the Spanish mainland enclave of Rio Muni, they continued to serve as exclusive plantations for the production of tropical crops such as cocoa, coffee, sugar and bananas for the metropolitan powers, who made no secret of their intention to keep them for ever as colonial possessions. No attempt was made to provide them with modern liberal political institutions, and they remained right up until their recent attainment of independence as a preserve of absentee landlords based in Europe and employing indentured or immigrant labour drawn from neighbouring countries on the African mainland, especially Eastern Nigeria.

EMERGENCE OF INDEPENDENT STATES

The end of European colonial rule came suddenly in French Equatorial Africa and, in the case of the Belgian Congo, quite unexpectedly. French Cameroons, whose status as a Trust Territory made it a focus of special international interest, was the first country in Central Africa to achieve its independence on 1 January 1960, to be followed on 30 June 1960 by the Belgian Congo. The four colonies of French Equatorial Africa proper, starting with Chad, followed in quick succession in August of the same year at intervals of two days in the following order; Chad on 11 August, the Central African Republic on 13 August, the Congo Republic on 15 August, and Gabon on 17 August. A year later, on 1 October 1961, as a result of a referendum organised under United Nations auspices, the southern part of British Cameroon, which had previously been administered with Nigeria, also became independent and opted to join with French-speaking Cameroon to form a federal republic.

The political and social forces immediately responsible for transforming the countries of French Equatorial Africa from colonial to independent status were identical in all essential respects with those which brought the former colonies of French West Africa to independence during the same period. Like their West African counterparts, the colonies of French Equatorial Africa initially accepted General de Gaulle's historic offer of 1958 to become self-governing, rather than fully independent states, within a Community or *Communauté* made up of France and her overseas territories. However, in 1960, as the African independence movement got fully under way, they finally decided to become independent republics in their own right but without severing their long-standing historical and cultural ties with France.

In the Belgian Congo things were entirely different both as regards the country's preparedness for independence and the attitude of the Belgians to the idea of handing over power to the Congolese. Up to 1958 the Congo formed a world of its own deliberately cut off by the Belgians from the political and social trends in the rest of Black Africa. Even within the country itself everything possible was done to discourage active contact among people from different regions and tribal backgrounds so as to reduce the opportunities for joint political action directed against Belgian rule. However, in 1958, sensing the new political mood in Africa, the Belgians for the first time allowed a number of Congolese leaders to travel outside the country and to establish contact with other African leaders by participating in various international activities, most notably the All African Peoples' Conference held in Accra.

The effect of these experiences was dramatic. The Congolese immediately began to make demands for increased participation in the country's political affairs with a courage and persistence which were quite novel. The Belgians saw the writing on the wall, especially since the French territories just across the Congo River were soon to become independent, and agreed to make a number of far-reaching changes aimed at eventually giving full independence to the Congo (Legum 1965). But things had been left too late; before the promised changes could be effected by the Belgians, they were forced as a result of mounting pressure by the Congolese and of disorders in various parts of the country, especially in the capital, Leopoldville (now Kinshasa), to hand over power to the Congolese on 30 June 1960.

The Belgians had done very little during the period of their rule in the Congo to prepare the Congolese for the assumption of political power; and before long there was a complete breakdown of law and order throughout the country, which later led to a civil war and the eventual secession of the rich Katanga province. Practically all the Belgian officials and experts who had been retained in order to assist in effecting a smooth changeover to African rule were forced to flee the country and the new Congolese government was obliged to call in United Nations troops and technical personnel to help with the restoration of order. But it took several years for things to return to normal, and for a long time Katanga was able to maintain its secession from the rest of the country with the support of certain powerful outside interests, including Belgium.

These tragic developments were due not only to the failure of the Belgians to give the Congolese the necessary political tutelage before the actual granting of independence; they were also to a large extent a direct consequence of the Congo's enormous size and the unusually wide tribal and ethnic diversity of its population, which made the task of holding the entire country together within a single administrative framework extremely difficult.

In the light of their experiences in the Congo, the Belgians were understandably reluctant to grant independence too suddenly to their Trust Territory of Ruanda-Urundi, especially since with the approach of independence the age-old antagonism between the two major ethnic groups, the Tutsi and the Hutu, was beginning to show signs of erupting into open conflict. However, as a result of intense pressure from the United Nations and from the other independent African states the territory was at last granted independence on 1 July 1962 and split into two states, Rwanda and Burundi.

SPECIAL PROBLEMS OF EMERGENT NATIONHOOD

Owing to certain special characteristics in the geography of Central Africa and in the policies followed by the colonial powers in control of the region the emergence of independent states proved to be a somewhat more complex process than in the rest of lowland Africa. As we have already seen, whereas in the French parts of the region the political and social climate were generally favourable, the economic base was relatively weak and inadequate. In the Belgian Congo, on the other hand, although the Belgians had succeeded in establishing an impressive economic infrastructure by the time of independence, very little had been accomplished as far as political institutions and manpower development were concerned. In the areas previously under Spanish and Portuguese rule the position was even more unsatisfactory because both the economic and political aspects of development had been woefully neglected and the territories concerned formed little more than tropical estates under the control of absentee landlords mostly resident in the metropolitan countries.

Because of the geographical heterogeneity of Central Africa it is impossible to discuss these problems in a general way, and the best approach is to give a separate account of the special problems facing each of the major territorial divisions.

THE FORMER FRENCH EQUATORIAL AFRICA AND CAMEROON

Historically and politically, the heart of this empire stretching from the Congo River to the Tibesti Highlands along the northern border of the Republic of Chad was Gabon. After Senegal, Gabon was the earliest French settlement on the African coast, and the people of Libreville, now the capital of Gabon, became French subjects as far back as 1893.

Because of its great latitudinal extent, the area has a very wide diversity of physical environments and an even greater variety of ethnic types. Understandably, the parts that received most attention politically and economically were those in the more attractive forest and savanna environments with relatively easy access to the coast, namely, the Congo Republic, Gabon and Cameroon, whilst the least developed and most remote was Chad. The Central African Republic, although it does not have an ocean outlet, had river access along the Oubangui and the Congo to Brazzaville, from where there was a rail link with Pointe Noire in the Congo Republic. Cameroon was perhaps the most fortunate state of all because under the Germans it had received a great deal of attention, especially in its coastal sections where one of the territory's more important railways had been built between the port of Douala and Yaoundé.

In terms of natural resources, however, the best-endowed country is Gabon, whose wealth in such important minerals as manganese, high-grade iron ore, uranium and petroleum gives it a valuable basis for industrial development. In addition to minerals the country is also rich in timber and has the potential for the production of several tropical crops. But it suffers from one serious handicap – the absence of an adequate labour supply, which is a direct reflection of its low population.

145

Indeed, it can be said that apart from the lack of adequate communications the main obstacle to economic development in the whole of this part of Central Africa is the sparseness of the population. The country with the highest population density is Cameroon, which has an average of 13.5 persons per sq. km (34.8 persons per sq. mile), followed by the Congo Republic with an average of 3.9 persons per sq. km (10 persons per sq. mile), and then Chad, the Central African Empire and Gabon, with densities of respectively 3.2, 2.8 and 2 persons per sq. km (8.1, 7.2 and 5 persons per sq. mile) (Fig. 34, p. 255).

Another general drawback of the area is the mountainous and highly accidented nature of the terrain, which consists essentially of a series of river divides. This makes the construction of railways and roads extremely difficult and expensive and is responsible for the isolation suffered by large portions of the interior. However, the resulting inconveniences are to some extent offset by the fact that the combination of mountains and high rainfall in the equatorial sections has given the region a very considerable hydro-electricity potential, which has been used to good effect, particularly in Cameroon, where the Sanaga River has been dammed to provide electricity at Edea for tropical Africa's first aluminium smelter as well as for domestic and other industrial uses in Douala and Yaoundé.

Although all these countries have several things in common as a result of their physical characteristics and the fact that all of them except the southern portion of the former British Cameroon were under French rule until they achieved independence, there is an important distinction between Cameroon and the other states. In the first place, Cameroon began as a German colony and then became a Mandated Territory, following Germany's defeat in the First World War, and subsequently a Trust Territory under the United Nations. It has therefore lacked administrative continuity and has tended to receive rather less attention from France than the other countries of the region. To complicate matters further, the recent addition of the southern section of the former British Cameroon to the French portion has introduced an English-speaking element with somewhat different political and economic traditions which now has to be grafted after forty years of enforced political separation to the French-speaking section of the country. This has added a serious complication to the difficult task of nation-building which faces the country and imposed upon it the intricate problems inherent in the development and administration of a bilingual state.

ZAIRE

In size as well as in terms of population, Zaire dominates the rest of Central Africa. Next to the Sudan, it is the largest country in the whole of Africa and perhaps the best endowed in terms of natural resources and economic potential. But it has probably had a more turbulent history since independence than any other country, although since 1969 it has enjoyed a considerable measure of political stability.

As we have already seen, the granting of independence to the country by Belgium in 1960 was immediately followed by widespread violence which finally culminated in the secession of the rich Katanga Province under Moise Tshombe and the murder of the country's first Prime Minister, Patrice Lumumba. United Nations troops and experts were called in to restore order and stayed in the country from August 1960 to June 1964. By the time they left the secession of Katanga had been ended and Tshombe was Prime Minister of the whole country. However, within three months, fresh troubles broke out in Stanleyville in the eastern part of the country, where a group of leftist rebels set up a 'People's Republic' and proceeded to murder scores of white hostages and thousands of Congolese. It took the intervention of Belgian para-troopers to restore order after nearly a year of bitter fighting in which white mercenaries employed by the Congolese government played a notable part. But fresh troubles broke out again in 1967 as a result of an attempt by Belgian and French mercenaries with the backing of dissident local forces to take over the eastern part of the country, and it was not until after two years that peace was restored.

In the meantime an intense struggle for power had been going on among the various political leaders in the country which led to the creation of several rival factions, until eventually on 25 November 1965 the head of the army, General Joseph Mobutu, ousted both the President and the Prime Minister and made himself President. Shortly afterwards a determined drive began to obliterate all vestiges of the country's colonial connection with Belgium and several place-names were changed: Leopoldville was renamed Kinshasa; Stanleyville, Kisangani; and Elisabethville, Lubumbashi. The name of the country itself, which at the time of independence was Congo Leopoldville, to distinguish it from Congo Brazzaville on the other side of the river, was changed to the Democratic Republic of Congo, and once again (in October 1971) to its present name, Zaire, which was the local name applied to the Congo River by the indigenous inhabitants at the time the Portuguese first arrived in the region.

That Zaire should have been faced with such formidable problems after the achievement of independence is not surprising, considering its huge area, which covers 2,345,409 sq. km (over 900,000 sq. miles), and the fact that it contains some 250 different tribal groups, which at the time of independence had still not acquired a sense of belonging to a single political entity. A federal form of government might have been a useful first step towards the achievement of this objective; but the strong separatist tendencies which manifested themselves immediately upon the termination of Belgian rule made such a proposition highly risky and argued for a strong centralised government such as the country now has. But even under the present authoritarian government the danger of centrifugal tendencies reasserting themselves still remains a serious potential threat to political stability, and it will be some time before the country achieves the stamp of a true nation state.

Without any doubt, Zaire's strongest asset is its enormous economic potential. It contains the largest area of equatorial and tropical forest to be found in any African country and in this respect invites comparison with the Amazon area of Brazil in South America. It is immensely rich in such tropical products as rubber, timber and ivory, as well as in a host of tropical crops like cocoa, bananas, coffee, cotton and palm oil. But its real economic strength lies in its exceptional mineral wealth and the enormous hydro-electric power potential of the Congo (Zaire) River. Most of the mineral resources are concentrated in the south and east, but diamonds are found in the basin of the Kasai River in the south-west. Zaire is the sixth largest world producer of copper, which yields more than 75 per cent of the country's export income. It is also the world's leading producer of cobalt and industrial diamonds, the eighth largest producer of manganese, the ninth largest producer of tin and the tenth largest producer of gold. Other mineral products include silver, zinc, cadmium, germanium, lead, coal and uranium.

Although its large size is one of Zaire's principal assets, it also constitutes a handicap because effective development and political control are impossible without effective transport and communications. This fact was recognised early in the country's modern history by King Leopold II, who was responsible for the famous dictum: 'coloniser c'est transporter'.

Large sections of the Congo River and its principal tributaries, the Kasai and Oubangui, are navigable, but continuous navigation is rendered impossible by the presence of numerous rapids and falls throughout the Congo basin. In order to develop the country, railways are required to supplement the navigable waterways, hence Stanley's observation that without railroads the Congo is not worth a penny. Although this was clearly an exaggeration, it nevertheless served to underline the importance of transportation for the country's development.

One of the most outstanding achievements of Belgian rule was the excellent system of communications which was laid down in the country before independence. By linking the navigable sections of the Congo (Zaire) and Kasai Rivers with railways the Belgians created a giant network of communications extending practically over the entire country (Fig. 19). Supplementing the waterways and railways is an

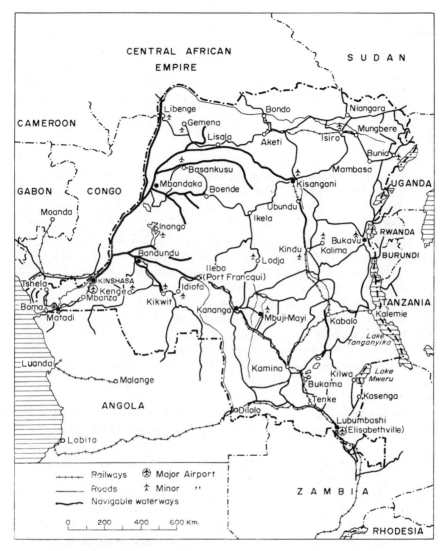

Fig. 19. Rail and river communications in Zaire

efficient system of air routes, which have done more than any other form of transport to bring the different parts of the country together within easy and rapid communication with the capital, Kinshasa. Roads have also played an important part in the development of Zaire; but owing to the enormous distances between the main centres of activity their role has been far more localised and strictly ancillary to rail, water and air transport.

Despite its excellent system of internal communications, Zaire is

149

severely restricted in its outlets to the sea. Below the port of Matadi the whole of the south bank of the Congo River is controlled by Angola. On the north, only a small area bordering the mouth of the river is under the exclusive control of Zaire; but even this is further constricted by the presence of the former Portuguese enclave of Cabinda. It is noteworthy that in 1927 the Belgians had to exchange no less than 3,500 sq. km (1,350 sq. miles) of the country for 3 sq. km (1.2 sq. miles) of Angolan territory behind the port of Matadi in order to enable vital extensions to be made to the port and its railway yard.

Given political stability, there is little doubt that Zaire can look forward to a bright economic future, thanks to the physical infra-structure laid down by Belgium while it was in control and the country's own inherent wealth in natural resources. The prolonged and bitter internal conflicts which erupted immediately after the achievement of independence showed in a dramatic fashion that an important pre-requisite for political stability was the conscious direction of all parts of the country towards a single focal point and the neutralising of all possible centrifugal forces. In short, there seemed to be an urgent need for the reorganisation of the administrative structure so that no single unit or tribal group would feel itself strong enough, as had happened in the case of Katanga and the eastern part of the country, to challenge the authority of the central government. At the same time, however, there was the danger that the existence of a multiplicity of administrative units could serve to defeat and undermine the all-important objective of national unity.

The search for an effective answer to these difficult problems has been among the principal preoccupations of the government ever since the attainment of independence. In 1961–62 the country's six original provinces were further subdivided, unfortunately as it happened, largely along tribal lines. Leopoldville province was divided into three, Kivu into three, Kasai into five, and Katanga into three, thus pro-ducing the unwieldy total of twenty-one units for the whole country. In terms of the country's area this scheme seemed reasonable enough, but considering the fact that the total population at the time was little more than 15 million the resulting units turned out to be too small in terms of population to be administratively and economically viable. Since then, a further reorganisation has taken place, and the country is now divided into the eight administrative provinces (Fig. 20).

Certainly, this new scheme has the advantage of greater simplicity; but only time can tell whether it will be better able than the previous one to assist in promoting a greater sense of national unity.

Apart from the tribal factor, another major cause of disunity in the country in the early days of independence was the enormous economic power wielded by the rich foreign companies operating in the country,

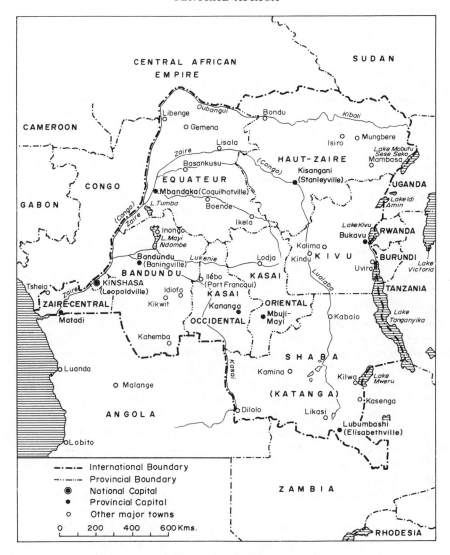

Fig. 20. Zaire: internal administrative divisions

Province	Capital
Zaire-Central	Matadi
Bandundu	Bandundu
Equateur	Mbandaka
Haut-Zaire	Kisangani
Kivu	Bukavu
Kasai Oriental	Mbuji-Mayi
Kasai Occidental	Kananga
Shaba	Lubumbashi

especially in mining, agriculture and transport, and their consequent ability to support separatist movements for reasons of self-interest.

By steering a middle course between the forces bent upon a policy of immediate nationalisation and those favouring a policy of complete *laissez-faire* towards these foreign commercial and financial empires, the present government of Zaire has been able to maintain international confidence in the country's economic affairs, while ensuring an increasing share of the profits of these enterprises for the benefit of the country. Admittedly, a great deal more remains to be done to make the people of Zaire masters of their own house, but the start so far made is clearly promising and augurs well for the future provided sufficiently vigorous steps are taken to correct the social deficiencies of Belgian administration, especially in the fields of higher education and manpower development.

RWANDA AND BURUNDI

The two kingdoms of Rwanda and Burundi, which prior to their achievement of independence were administered together by Belgium as the Trust Territory of Ruanda-Urundi, date back to the sixteenth century or so, when they were established by the tall cattle-keeping Tutsi (sometimes referred to as the Watutsi or Batutsi) who moved in from Ethiopia. The Tutsi, who now form only about 9 per cent of the population in Rwanda and 14 per cent in Burundi, imposed themselves on the indigenous Bantu peoples known as the Hutu or Bahutu who form about 85 per cent of the population in both countries. Under Belgian rule the Tutsi rulers were supported by the administration and were thus able to maintain their aristocratic position, but the hostile relations between them and the Hutu remained. Exactly why the Belgians supported the Tutsi rulers is not clear, since in the Belgian Congo their policy towards the local chiefs was to diminish or neutralise their power and rule through their own appointed functionaries.

With independence the two kingdoms became separate states and more or less reverted to their former geographical isolation as small mountain states located right in the heart of the continent and without any outlets to the sea. The Belgians sought to introduce administrative reforms aimed at giving the Hutu a greater share in the management of affairs shortly before the countries became independent, but the attempt came too late to have any effect. Thus, when independence came the Tutsi still retained their privileged position and the hostility between them and the Hutu remained unchanged.

It was in Rwanda that this hostility first broke out into open conflict. Just before the grant of independence, a strong Hutu political party emerged which enjoyed the support of the Belgian rulers, and in the elections which preceded the actual transfer of power by Belgium this party swept the polls and the people in a national referendum decided to abolish the office of Mwami or Head of State, which for centuries had been a Tutsi preserve. This led to an armed attack against the Hutu by the dissident Tutsi, many of whom had earlier fled into the neighbouring countries. Although the attack was successfully repulsed, it led to the subsequent massacre of thousands of Tutsi still resident in the country. The numbers killed have been estimated at between 1,000 and 14,000. As a result of these massacres large numbers of Tutsi again fled the country and took refuge in specially organised camps in neighbouring states, mostly in Zaire, Uganda, Burundi and Tanzania, much to the embarrassment of the Rwanda government which sees them as a potential danger and threat to itself.

In the initial post-independence period Burundi also faced similar political problems arising from the traditional hostility between the two peoples. As in Rwanda, a Hutu movement also emerged here just before independence and was able to win political control. However, the Tutsi rulers were more politically astute, and by tactfully adopting a non-partisan attitude in politics were able to retain the office of Mwami in Tutsi hands, while leaving the actual government of the country in the hands of the Hutu, who continued generally to enjoy the support and loyalty of the Tutsi aristocracy. All is not well, however, for the Tutsi have since succeeded in regaining their former political control through a number of military coups.

Both Rwanda and Burundi are poor countries with shaky economies based on the production of coffee, tea, cotton, cattle and tin (in the case of Rwanda) which are their principal export commodities. Their poverty is further aggravated by their high population density, which averages between 135 and 160 persons per sq. km (350–412 persons per sq. mile), the highest on the African mainland (Fig. 34, p. 255). As a result of their limited economic base and their enormous population pressure, many people from these two countries emigrate into the neighbouring countries in search of employment. It is believed that over 300,000 emigrants, mostly from Rwanda whose political instability has created large numbers of refugees, live in Uganda, which immediately adjoins it and with which it has close ethnic affinities; but most of the emigrants from Burundi go to Tanzania and Zaire.

Apart from the paucity of their economic resources, the highly mountainous nature of these two countries and their location far in the heart of the African continent make both internal and external communications, especially the latter, extremely difficult, although on the credit side Rwanda's relief offers ideal conditions for the generation of hydro-electricity. The two countries are separated by distances of over 1,208 and 1,932 km (750 miles and 1,200 miles) from the Indian and

Atlantic Oceans. Neither country has any railways, and their principal means of internal communications are by road, while air transport offers the only means of direct, uninterrupted communication with the outside world from the airport located at Kigali, the capital of Rwanda. However, because of its position on Lake Tanganyika, Burundi has water communications between its capital, Bujumbura, and the Tanzanian lake port and rail terminus of Kigoma. For their trade outlets to the outside world the two countries are mostly dependent on Dar es Salaam in Tanzania and Mombasa in Kenya.

As in the case of Zaire, manpower development was seriously neglected during the period of colonial rule under Belgium, a fact which has greatly hampered the development of both countries. But far the greatest problem facing them today, especially Rwanda, is the failure since independence to establish a reasonable degree of political stability on account of the deep-seated hostility between the two principal tribal and ethnic groups. Coupled with this are the problems posed by their very slender economic resources and their excessively high population densities. Thus, despite their attractive mountain environment, they are obliged to depend on what is hardly more than a subsistence economy with very limited prospects for future improvement.

THE FORMER PORTUGUESE AND SPANISH TERRITORIES

Apart from the mainland enclave of Rio Muni and its tiny estuarine islets of Corisco and the Elobeys this portion of Central Africa consists of four small volcanic islands arranged in a north-east to south-west direction along a major fault-line whose course on the mainland is marked by Mount Cameroon, which is still an active volcano, as well as by the Cameroon and Adamawa Highlands. The largest of the islands is Fernando Po, with an area of 2,035 sq. km (785 sq. miles), followed in order by São Tomé (855 sq. km or 330 sq. miles) Principe (109 sq. km or 42 sq. miles), and Annobon (17 sq. km or 6½ sq. miles), while Rio Muni together with Corisco and the Elobeys covers an area of 25,921 sq. km (slightly over 10,000 sq. miles).

Of the four volcanic islands, only Fernando Po was inhabited when the Portuguese first arrived in the area in the fifteenth century; and for a long time it was used by them and later by the Spanish, who acquired it together with Annobon from the Portuguese, as a major base for the shipment of slaves from the neighbouring African mainland. However, because of Fernando Po's strategic location in relation to the Nigerian coast, the British navy leased bases on the island between 1827 and 1843 for use in suppressing illegal slaving activities in the region after the abolition of the slave trade as well as for the settling of freed slaves. In 1900, as a result of the Treaty of Paris which formally determined the boundaries of Rio Muni, the Spanish, who had previously held only the coastal portions of the territory, acquired a large area in the interior and thus substantially increased their territorial holdings in this part of the continent.

Small as the four Portuguese and Spanish volcanic islands were, they assumed considerable economic importance during the period of the slave trade, but especially after its abolition, as a source of valuable tropical crops grown on European-owned plantations. In the initial stages the plantations employed slave labour; but after the abolition of slavery they depended mainly on indentured and immigrant labour derived from nearby territories on the mainland, and sometimes even on forced labour pressed into service from Angola and Rio Muni. But the ownership and management of the plantations continued to remain in the hands of absentee European landlords mostly resident in the metropolitan countries.

Economically, Fernando Po is far the most important of the islands. It produces a large variety of tropical crops, but the one for which it is best known is cocoa, the quality of which is among the highest in the world even though the actual quantity involved is only between 30,000 and 40,000 tons per annum. Fernando Po is literally the seed-bed of the present cocoa industry of West Africa. It was from here that the seedlings which laid the foundations of Ghana's cocoa industry were smuggled into the Gold Coast in 1879 by Tetteh Quashie, a native of the country who had worked as a labourer in the cocoa plantations of the island. But Fernando Po is so heavily dependent on foreign labour that as much as 75 per cent of its present population are believed to consist of immigrants from Eastern Nigeria, including several thousands of refugees who fled there as a result of the Nigerian civil war.

As is to be expected, the assimilation of these large numbers of immigrants from countries with markedly different political and cultural backgrounds to the indigenous populations of the island has created serious social and political problems, and can be said to have been the main cause of the political upheavals that led to the granting of independence to her territories by Spain on 12 October 1968, followed by Portugal in 1975.

Since independence these problems have been further compounded by internal conflicts arising from the political fusion of the two islands of Fernando Po and Annobon (now renamed as Macias Nguema Biyogo and Pigalu, respectively) whose inhabitants consist mostly of Spanish settlers and people of mixed racial origin with a long exposure to European ways with the mainland territory of Rio Muni and the small coastal islets of Corisco and the Elobeys inhabited mostly by the Feng, a Bantu people who were subjected to ruthless exploitation under Spanish rule. Although the Feng are not as sophisticated as their island compatriots, they are far more numerous and now form the politically dominant group. Also, the country's economy has been seriously undermined by the withdrawal since 1970 of most of the Nigerian immigrant population of some 60,000, who previously formed the main source of labour on the cocoa plantations on Fernando Po, on account of the oppressive policies of the present government, while for the same reason practically all the European population and large numbers of the indigenous population from both Fernando Po and Rio Muni, variously estimated at between 12,000 and 100,000, have fled from the country and sought refuge in Spain, Gabon and Cameroon (Legum 1974 and 1975). In these circumstances the country has become increasingly alienated from its neighbours, especially Gabon with which it has had serious boundary disputes, without in any way offering the inhabitants a better life than they enjoyed under Spanish rule. More than anything else, there would seem to be a clear need for national reconciliation and the adoption of measures aimed at the political consolidation of the various parts of the country which under colonial rule were administered as separate units whose only common denominator was their political and economic subjugation to Spain.

By comparison with Equatorial Guinea, the problems facing the two former Portuguese islands of São Tomé and Principe, which achieved their independence on 12 July 1975, appear to be more simple. However, like their Spanish neighbours, now that these islands are independent, they must stand on their own feet despite their very limited economic base and their relative lack of any serious political experience as a result of the long period of oppressive and paternalistic rule to which they were subjected under Portugal.

How these two new states fare in the course of the next few years will depend to a large extent on the sort of political and economic relations which they are able to establish both with their former rulers and with their African neighbours on the mainland with whom they have had very few real contacts over the past five centuries.

8

Southern Africa

Southern Africa comprises approximately the whole of the continent to the south of the Republic of Zaire and bordered on the east and west respectively by the Indian and Atlantic Oceans. As shown in the table below, it consists of ten political divisions ranging in size from Swaziland with an area of 17,363 sq. km (6,704 sq. miles) and a population of 493,000 to the Republic of South Africa and Angola with areas of 1,221,037 and 1,246,700 sq. km (471,443 and 481,351 sq. miles) and populations of 25,471,000 and 5,800,000 respectively (Fig. 21).

Political divisions	Area Sq. km	Sq. miles	Population 1975 or latest
Angola	1,246,700	481,351	5,800,000
Botswana	600,372	231,804	677,000
Lesotho	30,355	11,720	1,038,500
Malawi	118,484	45,720	5,044,000
Mozambique	783,030	302,328	9,239,000
Namibia	823,169	317,827	746,328
Rhodesia	390,236	150,670	6,100,000
South Africa	1,221,037	471,443	25,471,000
Swaziland	17,363	6,704	493,728
Zambia	752,614	290,584	4,896,000
Total	5,983,360	2,310,178	59,505,556

In relation to the rest of the continent, the most distinctive features of the region are its relatively narrow width which renders it almost peninsular in character, its high and extensive plateau elevations averaging more than 1,000 m above sea level, and, most notably of all, the fact that permanent white settlement is more extensive and far more deeply entrenched here than in any other part of the continent, with the result that over a long period of time racial conflict between Africans and Europeans has overshadowed practically all other considerations in the political geography of the region and produced social tensions without a parallel in the rest of the continent.

The conflict is not simply the result of a clash of political and economic interests between the different racial groups. As has been shown increasingly by events in the Republic of South Africa,

Fig. 21. Southern Africa: political divisions and railways

Rhodesia, and Namibia, which are today the only remaining strong-holds of white domination in the continent, it stems from a deeply held conviction on the part of the white minorities that they are racially superior to the millions of Africans and other non-Europeans who live both within the region and in the rest of the continent.

In an age when doctrines of racial superiority have not only been debunked but are widely condemned by the entire international com-munity, it is not surprising that Southern Africa, and especially South Africa and Rhodesia, has become a focus of intense world interest and concern and for the peoples of Black Africa in particular the ultimate symbol of colonial oppression.

EUROPEAN COLONISATION AND THE POLITICAL EVOLUTION OF SOUTH AFRICA

With the exception of the former Portuguese colonies of Angola and Mozambique and the former British colony of Nyasaland (now

Malawi), which were penetrated directly by Europeans from the immediate or adjoining coastal areas, European contacts with Southern Africa took place from the southern point of the continent, at the Cape of Good Hope, starting with a small Dutch settlement established on the shores of Table Bay in 1652 to serve as a victualling station for ships sailing between Europe and the Dutch possessions in the Indian Ocean. Earlier, as we have seen in the preceding chapter, the Portuguese had established footholds along the northern section of the Angolan coast, but hardly any attempts at colonisation had occurred farther south on account of the generally inhospitable nature of the coast, which for a stretch of some 1,600 km (1,000 miles) between the southern border of Angola and the mouth of the Orange River in the territory now known as Namibia is marked by the Namib desert.

The Cape of Good Hope, more popularly known simply as the Cape, was ideal for settlement from many points of view. Firstly, Table Bay, which adjoined it, offered an excellent natural harbour for ships. Secondly, unlike the dry coastal strip to its north and the hot, humid coast of Natal on the other side of the continent which is washed by the warm waters of the Agulhas current, it lay within an area with an attractive Mediterranean climate. Thirdly, the landforms consist of a series of relatively low east–west ranges, the Swartberge and the Langeberge Mountains and other related ranges, with two intervening plateaux known as the Little and Great Karoos, which led like steps to the main plateau of Southern Africa, thus giving relatively easy access to the interior through a series of natural passes or 'poorts' (Fig. 22).

It is known that before the arrival of the Dutch the Portuguese had tried to establish a settlement at the Cape but were obliged to abandon the idea, following the massacre of the first party which they sent ashore by the local inhabitants.

The original idea of the Dutch in establishing their victualling station at the Cape was in no way to found a settlement for the future colonisation of the region, and the orders of Jan van Riebeeck, who was entrusted with the task, were simply to build a fort large enough to hold about a hundred soldiers and such sick sailors as might be left by the ships of the Dutch East India Company sailing to the Indies. In time, however, the station developed into a settlement for retired servants of the Company, who grew crops and reared cattle for passing ships, and by 1795 had expanded to some 15,000 colonists. But in keeping with the original purpose of the colony severe restrictions continued to be imposed on the movements and activities of the settlers (Wellington 1955).

Right from the outset one of the most thorny problems that faced the Dutch authorities at the Cape was the question of the relations between the colonists and the local Hottentot inhabitants, who were essentially

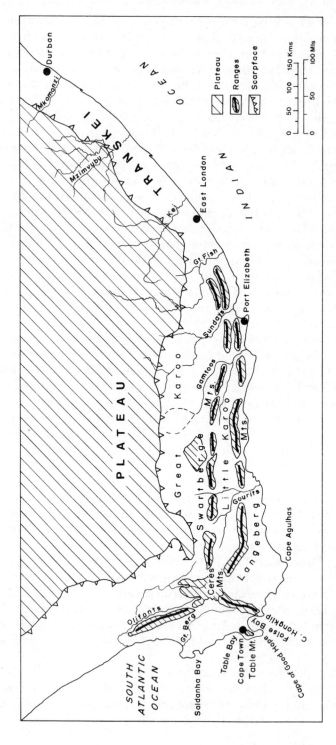

Fig. 22. The Cape folded ranges and the plateau of South Africa (greatly simplified). Note the gaps or 'poorts' in the ranges

a pastoral people rearing sheep and cattle which they bartered for manufactured goods from the Dutch. However, the circumstances of the settlement, which initially consisted entirely of males, led inevitably to a considerable degree of social intermingling with the Hottentots, thus giving rise to a mixed stock of the two races known as the Cape Coloureds to whom were added in due course further admixtures derived from slaves brought in by the Dutch from West and East Africa. But relations between the Dutch and the Hottentots were never really easy and increasingly deteriorated as a result of incessant accusations of trespass on either side and the realisation by the Hottentots that the Dutch intended to settle permanently in the region.

The efforts of the authorities to restrict the settlers to the confines of the initial colony in order to avert conflict with the Hottentots proved futile, and very soon the settlers began to move eastwards from the Cape along the coast and into the fertile valleys within the coastal ranges south of the main plateau escarpment. In the process they began to encounter various Bantu peoples moving in the opposite direction into the fertile lowlands of Natal, and thus began the first really serious border skirmishes between the Dutch settlers and the Bantus, culminating in the battle of the Great Fish River in 1779.

The next important development in the political evolution of Southern Africa took place during the Napoleonic wars, when the British took over the Cape from the Dutch in order to protect their sea route to India and subsequently retained the territory as a British colony. With the arrival of the British, the restrictions upon the Dutch in their dealings with *men of colour*, as they called the Black inhabitants of the region, steadily increased, making the Dutch even more resentful of the British presence.

Between 1685 and 1688, long before the British arrived, a number of Dutch orphan girls had been settled at the Cape to maintain domestic life, followed by a sizeable number of French Huguenot refugees from the Charente region of France, who brought new skills in the arts of viticulture and wine making, but these became quickly assimilated with the Dutch settlers. The arrival of the British, however, introduced a new and distinctive element into the population, which was reinforced in 1820 by some 5,000 British immigrants who were settled around Port Elizabeth, just west of the Great Fish River, to serve as a human buffer between the Dutch and the Bantus. However, because the land here did not particularly favour the kind of agriculture familiar to most of these settlers they quickly drifted into the towns or turned to trading among the Bantus east of the river.

Expansion of the British community and the increasing efforts of the government to regulate relations between whites and blacks at the Cape and put them on a more equitable and Christian basis made the Dutch settlers even more resentful of the British. Finally, under their leader, Piet Retief, they decided to turn their backs on the Cape and seek a new life of their own on the more open and freer lands of the wide plateau region in the interior, where they could escape from further restrictions and preserve what they regarded as 'proper relations between master and servant'. Here, the prevailing geographical conditions of the dry veld turned them from an agricultural people into a pastoral people and further deepened their isolation from the rest of the country and the world.

By 1836 the drift away from the coast had gathered momentum and developed into a massive movement known as the Great Trek. The Dutch settlers engaged in the movement saw themselves as a chosen race charged with a divine mission to seek their freedom in a new and hostile land inhabited by black heathens whom they regarded with the utmost contempt and hatred. By the middle of the century the Great Trek had resulted in the establishment of two virtually independent Dutch republics, the Transvaal and the Orange Free State, while the Cape Colony and Natal remained fully under British control.

The establishment of the two republics involved several fierce and bitter wars between the Dutch and various Bantu peoples who under such leaders as Chaka and Dingaan proved to be formidable adversaries. Although the Bantus were mostly defeated, because of their large numbers it proved impossible to eliminate them completely, and substantial pockets of them remained in various parts of the country, especially in Natal and the eastern parts of the Cape Province, where they form today the greatest concentrations in the so-called Native Reserves or Homelands, and also in Basutoland (now Lesotho), where some of the most intrepid Bantu peoples were pushed into the mountain fastnesses of the Drakensberg Mountains by the advancing Dutch trekkers.

But the Great Trek did not end the conflict between the British and the Boers, only about a third of whom had actually taken part in the exodus, leaving the greater majority within the Cape, where they continued to wrangle with the British authorities over the question of their rights as opposed to those of the non-white populations. Moreover, the British refused to recognise the claims of the new Boer republics to independence and insisted on regarding them as still subject to the authority of the British government and therefore amenable to British laws and codes of conduct in their dealings with the non-white races, even though with the steady expansion of the British settler population the earlier differences between the British and the Boers in their attitudes towards the non-white races, especially the Bantus, were becoming increasingly blurred.

What really brought the age-old animosity between the British and the Boers to a head, however, was the refusal of the Boers to allow the British a fair share of the wealth from the rich diamond and gold deposits which had been discovered in the Orange Free State and the Transvaal between 1867 and 1886, despite the fact that British capital and enterprise had been largely responsible for the exploration and exploitation of these deposits. Eventually, a major conflict – the Boer War of 1899–1902 – broke out between Great Britain and the two Boer republics of the Transvaal and the Orange Free State.

Although the Boers were defeated in the war, the final settlement took the form of a compromise arrangement whereby the Transvaal and the Orange Free State were reduced to the status of Crown Colonies until 31 May 1910, when under the Act of Union passed by the British parliament in 1909 they were joined with the Cape and Natal as Provinces within a self-governing Union of South Africa.

The net effect of the Act was to compromise the political rights of the African and other non-white populations in the Union, especially in the two former Boer republics, whose discriminatory laws debarring non-whites from the franchise were allowed to stand. The only group whose franchise rights were explicitly protected were the Cape non-whites, in particular the Cape Coloureds, who right from the start had been accorded special privileges by the whites on account of their European ancestry. Besides the Africans, other victims of racial discrimination in the Union were the large number of Indians whom the British had brought into the country towards the end of the nineteenth century to work as indentured labour on the sugarcane plantations in the humid coastal plains of Natal as well as the much smaller number of Chinese and other Asians who had entered the country on their own to work as labourers in the newly discovered diamond and gold fields on the plateau. Indeed, in the case of the Chinese, such was the hostility of the Boers towards them that the Orange Free State, which was the stronghold of Boer nationalism, refused outright to allow them even to set foot within its boundaries.

Thus, despite all the efforts of the British government and the previous Dutch administration at the Cape to establish harmonious and equitable relations between the different races in South Africa, very little had in fact been achieved, thanks largely to the uncompromising attitude of the Boer settlers. By the time the Union came into being, deep racial conflicts had developed not only between the whites and the non-whites, but also between the British and the Boers, who formed the majority of the white population.

The non-white population consisted of a number of distinct racial groups, of which the largest were the Bantus, who far outnumbered the rest, the Coloureds and the Indians. The Coloureds were mostly concentrated in the Cape Province, the Indians in Natal, while the Bantus were generally to be found in the warmer eastern and northern parts of the country. The Bantus did not form a single homogeneous people but rather an agglomeration of several distinct tribes, each with its own cultural traditions, although they all belonged to the same racial stock. However, because of their very large number and the initial challenge which they posed to white expansion it was chiefly against them that most of the discriminatory racial legislation in the country came to be directed, and in the face of white economic expansion into the more attractive agricultural lands and the mineral-rich areas they were steadily deprived of their lands and reduced, along with the other non-white groups, to the status of second-class citizens whose main role in the economic and social life of the country was to provide cheap labour for the white population.

Among the whites, the Boer element, which mostly comprised the rural and less-prosperous section, deeply resented the political and economic power of the British section and set about to reduce its influence and to bolster up Boer nationalism through the formation of political parties and the development of the Afrikaans language, which the Dutch settlers had evolved over the centuries into the dominant language of the Union.

Such a state of affairs could not possibly augur well for the future, and throughout its history South Africa has lived under the shadow of deep racial conflict which has grown increasingly bitter with the years and the decline of British influence in the conduct of the country's political affairs.

LAND POLICIES IN SOUTH AFRICA

Over and above all the discriminatory legislation to which the African population of South Africa has been subjected, far the most important and controversial issue has been that of land, owing to the fact that the Africans are predominantly an agricultural people directly dependent on the land for their livelihood.

In the early days of European settlement land was so plentiful around the Cape that its acquisition by the colonists did not present a serious problem for the Hottentots and Bushmen who constituted the indigenous population. Problems of land ownership really began to arise as the European frontier was extended far to the east and north in areas densely occupied by the advancing Bantu peoples.

To solve these problems the British government, which was then in control, decided to create *reserves* for natives after the middle of the eighteenth century. The two most progressive provinces in this respect were the Cape Province, especially east of the Great Fish River, and Natal. In the Transvaal, Boer farmers occupied land at their own option without any regard for the needs of the African population, although later on the government made some provision for the Africans. In the Orange Free State the native population had largely been exterminated or driven out through war, but some provision to accommodate them was made towards the end of the nineteenth century.

At the time of the establishment of the Union in 1910 Native Reserves totalled roughly 9.5 million ha (23.5 million acres) or approximately 7.13 per cent of the total area of the country. As part of the reconstruction which followed the Boer War, a Native Affairs Commission was appointed in 1903 to consider the official policy to be applied to natives throughout the Union, but with the main aim of protecting European interests by confining the right of land purchase by natives strictly to certain prescribed areas.

The Native Land Act which resulted from the recommendations of this Commission scheduled the areas already occupied by Africans and restricted further purchases by them, save in the Cape Province, where this prohibition was restrained by the courts. The Act proposed the setting-up of a Commission to recommend a permanent division of land in the country into African and non-African areas, and the Beaumont Commission which was consequently appointed for the purpose proposed the setting aside of 7.2 million ha (17.7 million acres) where Africans could purchase land, thus bringing the total area available to them to roughly 15.6 million ha (38.5 million acres) or approximately 13.3 per cent of the land surface of the Union.

The final decision on these proposals was left to each Province of the Union. Apart from the Cape all the Provinces reduced the area originally recommended, but no action was taken until 1936, when an Act passed for the purpose of implementing the allocation provided for the acquisition of a maximum of only 5.9 million ha (14.5 million acres) to be added to the land already scheduled for native occupation. A sum of £10 million was to be expended during the next ten years for effecting the necessary purchases, and a South African Native Trust with the governor-general as Trustee was charged with the responsibility of purchasing and controlling any lands acquired under this Act.

Not all land acquired by the Trust, however, represented a net addition to land previously occupied by natives, since some of the Crown land handed to the Trust already had Africans on it and several Africans on Crown land which did not fall within the allocation were treated as *squatters* and accordingly ejected. In the end, therefore, the result of these elaborate exercises was to grant the Africans a total area of 15,951,000 ha (39,417,000 acres) or roughly 13 per cent of the Union, mostly in the less-fertile parts of the country or on land which though initially fertile had become greatly depleted through overcropping

Fig. 23. Distribution of Native Reserves in South Africa

and overgrazing on account of the excessively high population density (Fig. 23).

Although the ostensible aim in establishing the Native Reserves was to provide the Africans with land for their exclusive use and fully protected against the possibility of alienation to the members of other races, there were several other reasons behind the arrangement which were aimed at promoting the economic and political interests of the European population.

In the first place, the creation of the reserves gave the European areas protection against possible encroachment by Africans, who were completely excluded from these areas except as wage earners and tenant-labourers working for Europeans. Secondly, by restricting the area of land available as reserves, the Europeans were able to force large numbers of Africans off the land who therefore formed a much more ready source of cheap labour for use by the Europeans than might have been the case if the Africans had been allowed to retain large areas for their subsistence. Thirdly, by deliberately scattering the reserves over a large area in relatively small units rather than large compact blocks the Europeans succeeded to a large extent in destroying the unity and solidarity of the African population. Only in the Transkeian territories and part of Zululand along the east coast were compact and homogeneous native blocks of land created, and that only because of the already existing concentrations of Africans in the area at the time of the demarcation. Finally, the reserves served as a convenient *shock absorber*, affording a place where aged and infirm Africans could retire after their years of useful employment outside without becoming a charge on the state.

Today the African reserves form some 264 separate units or Homelands in a sea of white lands. In 1970 there were nearly 15 million blacks in South Africa as compared with 2 million coloureds, 614,000 Asians and only a little over 3.5 million whites. Predominantly the Africans are farmers eking out a miserable existence from their grossly overpopulated and generally infertile lands; but because of the intolerable population pressure in the reserves and the great demand for labour on European farms and in the mines and factories located in the urban areas only about a third of them live permanently in the reserves, which even so are greatly overcrowded and quite incapable of supporting their populations. The highest concentration of African lands occurs in Natal, where they cover 33 per cent of the total area of the Province, followed by the Transvaal, the Cape Province and the Orange Free State with respectively 18.7 per cent, 9.2 per cent and 1 per cent.

In contrast, the European population of 3.5 million, which is mostly urbanised, have for their exclusive use an area amounting to about 87 per cent of the total area of the country, including the areas with the best agricultural lands and practically all the known mineral wealth.

For more than a century the political life of South Africa has been dominated by racial conflict made all the more intolerable by the fact that while it is recognised that the economic interests of the different races are closely interlinked every effort has been made by the Europeans, especially the conservative Boer elements, to enforce a rigid social separation between whites and non-whites.

Up to the time of the creation of the Union the British were in a position to maintain some semblance of justice in the relations between the Europeans and the rest of the population; but since then the Boers, who form the majority of the white population, have come to dominate the country's political life and have systematically shorn the non-whites of all their political rights and reduced them to a state of complete impotence by the enforcement of stringent racial laws.

THE POLICY OF APARTHEID

The hardening attitude of the Boer nationalists towards the Africans and other non-white races in South Africa assumed its most dramatic and vicious form in 1948, when the country's new Prime Minister, Dr Malan, summed up his government's racial policy as one of *apartheid*, an Afrikaans word meaning 'separateness'. Up to that time the non-whites had had a raw enough deal; but after the open declaration of the policy of apartheid and the many repressive measures which followed rapidly in its train life for the non-whites, and to some extent even for the more liberal elements of the white population, was suddenly transformed into an unbelievable nightmare of senseless oppression. In 1949 the Prohibition of Mixed Marriages Act was passed, making intermarriage between whites and non-whites a punishable offence; in 1950 the Group Areas Act, subsequently amended and consolidated in 1957, was passed, providing separate housing areas for the different racial groups, followed in 1953 by the Reservation of Separate Amenities Act prescribing separate transport facilities and other public amenities for the different races. Then in 1957 the most socially obnoxious law of all, the Immorality Act, was passed making all forms of sexual relations across racial barriers an offence (Legum 1965).

The intention behind all these laws and many others besides was manifestly clear: the South African government was determined to bring about a complete separation of the white and non-white populations based on the obnoxious doctrine of white racial superiority and a deliberate policy to prevent any kind of fusion among the different sections of the non-white population.

TRIBAL HOMELANDS

An aspect of the policy of apartheid which has aroused widespread controversy among the African population in recent years has been the decision taken by the South African government in the 1950s to create a number of tribal Homelands or Bantustans out of the existing 264 African reserves scattered throughout the country with the aim of encouraging the African population to withdraw from the white areas and establish semi-autonomous states of their own run along traditional lines under their own traditional rulers or tribal chiefs. Despite strong African opposition to the proposal and the many serious reservations expressed about its practicability by the Tomlinson Commission[1] specially appointed by the government to examine the

1 See summary of the Report contained in 'The Pattern of Race Policy in South Africa', published in the *Digest of South African Affairs*, April 1956, Government Printer, Pretoria. See also Legum (1965).

socio-economic implications of the proposal, the scheme was pushed through by the government early in the 1960s and a number of home-lands were set up, most notably in the Transkei and Zululand which have the most extensive areas of African reserves.

Although this was quite obvious even before the start of the experi-ment, experience so far has clearly shown that the Bantustan or Home-lands scheme cannot offer a satisfactory solution to South Africa's racial problems or serve in any way to demonstrate what its white advocates claim to be the more positive aspect of the policy of apartheid unless, as the report of the Tomlinson Commission pointed out, the basis of land allocation as between Europeans and Africans in the republic as a whole is drastically overhauled and considerable sums of money are spent on the improvement of the black areas in order to make them sufficiently viable economic units for accommodating and sup-porting the large influx of population envisaged for them under any thoroughgoing scheme of territorial division along racial lines. Besides, there is as yet no indication that any alternative means has been found for meeting the labour requirements of the Europeans, especially in the urban and industrial centres, which is the main reason for the presence of large numbers of Africans in the so-called European areas of the country.

In the tribal Homelands themselves the hopes of political autonomy held out to the African population have proved to be largely illusory since the overall control of affairs is in the hands of white commissioners-general appointed by the central government and only a small proportion of the local bodies responsible for the running of affairs consist of truly elected representatives of the people. In spite of this, some of the African tribal chiefs who have been appointed as heads of the Homelands have succeeded in using their positions to express fairly independent views – sometimes quite critical of the South African government – not only on the internal affairs of their own 'states' but on the racial policies of the republic as a whole. Although the Homelands admittedly represent an interesting experiment, in the present realities of South Africa there is very little likelihood that they can be developed to their logical conclusion or that even if this were to happen the majority of the African population would be prepared to accept such a blatantly inequitable territorial partition of the country as a final solution to the racial and political problems which have plagued South Africa for so many centuries. Most Africans, especially the younger generation, totally reject the idea of the Homelands and see them largely as irrelevant distractions created by the whites to divert their attention from the country's basic racial, economic and political issues.

In spite of this and of all the protests voiced against the homelands scheme by the International community and the Organisation of

African Unity, on 26 October 1976 the South African Government went ahead with its plans and declared the Transkei an 'independent' republic within the Republic of South Africa. The reaction of the United Nations was immediate. The General Assembly, which was then in session, voted overwhelmingly against any form of recognition of the Transkei and condemned the action of the South African government in imposing this meaningless form of independence upon the helpless inhabitants of the Transkei, the aim of which seemed nothing less than the perpetuation and entrenchment of the obnoxious policy of apartheid within the republic.

The expressed intention of the South African government is to establish altogether ten such Homelands. Apart from the Transkei, the other nine Bantustans will be: KwaZulu, Bophuthatswana, Lebowa, Gazankulu, Ciskei, Venda, Swazi, Basotho Qwa-Qwa and Ndebele.[2]

Whether or not these nine remaining Homelands will in fact be established as 'independent' republics remains to be seen; but in view of the world reaction to the Transkei and of the current political climate in Southern Africa it would be prudent for the South African government to give a second thought to the Homelands scheme, which clearly cannot assuage either local or international opposition to its apartheid policy.

EXTERNAL RELATIONS

Opposition to South Africa's racial policies has been widespread both within and outside the country, but very little in the form of positive results has been achieved. Determined efforts by a large section of the international community acting through the United Nations to bring South Africa to terms by the application of trade embargoes and other sanctions have been frustrated through lack of support from some of the major Western powers as well as by the economic resilience of South Africa itself.

The fact is that economically South Africa is one of the best-endowed countries in Africa and in the world. It is the world's leading producer of gold and diamonds, and has considerable reserves of coal, iron, copper and manganese to support its own industrial production, not to mention a number of other strategically important minerals such as uranium, chromium, tungsten and asbestos for which there is a great world demand (Fig. 24). It also has extensive areas of land located within different climatic zones for agricultural production for the supply of food and industrial raw materials. Besides, its location at the

---

2 'The South African Bantustan Programme: Its Domestic and International Implications', reprinted from *Objective: Justice*, Vol. 8, No. 1, Spring 1976, United Nations Office of Public Information, New York.

Fig. 24. Southern Africa: mineral resources and communications

southern tip of the African continent gives it a commanding position in relation to the sea routes between the Atlantic and the Indian Ocean; and the naval base of Simonstown, standing on False Bay, just south of Cape Town, was until recently a key stronghold in Britain's imperial defence system and is even now of considerable strategic interest to the West, especially in view of the growing naval strength of the Soviet Union in the Indian Ocean. Thus, despite international opposition to its racial policies it has been able to maintain a policy of complete defiance of outside opinion without endangering its own essential domestic needs or losing any of its important external trade.

In 1960, as a result of mounting hostility from the newly independent African and Asian states formerly under British rule which now form a majority within the Commonwealth family of nations, South Africa decided to become a republic and to leave the Commonwealth. Although it has since become a virtual outcast among the international community of nations, it still maintains strong trading ties with many of the more powerful nations both because of its economic assets and because of its strategic location at the southern tip of the African continent between the Indian and Atlantic Oceans.

In no area has South Africa's contempt of international opinion been more flagrant than in respect of its administration of the vast territory north of the Cape Province, formerly known as South West Africa but renamed Namibia since June 1968 by a decision of the United Nations.

South West Africa was formerly a German colony; but after the First World War it was handed over to South Africa as a Mandated Territory under the League of Nations. After the Second World War when the United Nations replaced the League of Nations the territory should properly have become a Trust Territory under the new world body, like all the former Mandated Territories of the League of Nations.

However, South Africa refused to recognise the rights of the United Nations over the territory and under the South West African Affairs (Amendment) Act of 1949 proceeded to incorporate it into its own administrative system in the face of world opposition and to bring it under its harsh racial laws. The renaming of South West Africa as Namibia by the United Nations in 1968 was intended to underline the fact that that body is still the rightful body for determining the territory's political future, and since then numerous efforts have been made by the United Nations to get South Africa to relinquish its claim to Namibia and grant the people their independence, but so far without any success.

There is no doubt that one of the main reasons underlying South Africa's reluctance to relinquish its hold of Namibia is the vast mineral wealth of the territory (Fig. 24). Since 1928, when diamonds were first discovered in the coastal gravels north of the mouth of the Orange River, Namibia has become the world's richest source of gem diamonds, yielding over £30 million annually. Subsequently other important minerals, such as copper, lead, zinc, vanadium, germanium, tin and several others have been discovered. But another and perhaps even more fundamental reason is that the territory serves as a useful buffer between South Africa and the rest of the continent, where African rule is now firmly established. For decades South African statesmen have sought to extend their country's sphere of influence in relation to the rest of the continent as far north as possible as a means of safeguarding white dominance within their boundaries and to ward off the creeping influence of black political ascendancy from the north. Thus, although the Limpopo River marks the northern boundary of the country, it was common for them in the inter-war period before the independence movement of Black Africa got under way to assert that a more meaningful boundary was the Zambezi River far in the heart of the continent.

Unfortunately for the white nationalists in South Africa this dream has not materialised, and with the attainment of independence by the former Portuguese colonies of Angola and Mozambique the boundary between South Africa and the independent states of Black Africa has been pushed well to the south of the Zambezi River. Even the Limpopo has been crossed; for the granting of independence to the former British protectorates of Basutoland (now Lesotho) and Swaziland has had the effect of thrusting far inside the boundaries of South Africa two exclaves of independent Africa, thus further undermining the insulation it has tried for years to secure against the advancing tide of black nationalism in the continent. In view of these developments it is not surprising that South Africa has so vehemently resisted the efforts of the United Nations to secure independence for Namibia which, apart from serving as a convenient political buffer, is also an important source of revenue to South Africa on account of its great mineral wealth.

It seems, however, that at long last South Africa has accepted the need for change in Namibia no doubt as a result of the continual pressure of world opinion but perhaps partly also because of the guerrilla activities against the government by the country's main nationalist movement, the South West African Peoples Organisation (SWAPO), which for some years now has been operating from bases in nearby African states. South Africa appears now to have accepted the principle of self-determination for the people of Namibia and has started constitutional talks with the various ethnic groups in the country as the first step towards the granting of full self-government. It is expected that this long-awaited event will occur some time in 1977 or 1978. However, it is highly unlikely that the form of government which South Africa has in mind for the territory will be acceptable to the nationalists, who have so far been excluded from the talks. Initial indications suggested that the type of government intended for Namibia and which the South African Government had described as 'ethnic democracy' will be based on the same communal principle which has been employed so successfully within the republic itself for dividing the various racial and ethnic groups and thus destroying the political effectiveness of the Africans who, here as in South Africa, constitute the overwhelming majority of the population – about 656,000 out of a total population of 746,000.

In South Africa itself the intensity of African opposition to white

oppression continues to mount. Since 1976 several spontaneous uprisings by Africans and other non-white groups have broken out in various
parts of the country, starting with the African township of Soweto on
the outskirts of Johannesburg, where demonstrations by schoolchildren
against the compulsory study of Afrikaans triggered off a series of
demonstrations in other parts of the country. Whatever the long-term
consequences of these events, they are of immediate political and
historical significance in themselves because they mark the first open
challenge by the African population on a nation-wide scale to the racial
policies of the South African government. Following this eruption, it is
certain that further acts of defiance can be expected in the future until
acceptable corrective measures are instituted.

South Africa is perhaps the best-armed state in the whole of Africa.
But there is ample historical evidence within and outside Africa to show
that the force of arms alone will not be sufficient to hold down unwilling
subjects and stem the tide of a popular mass uprising, especially where,
as in South Africa, the victims of oppression are not only in the majority
but also play a key role in the national economy as the principal source
of labour and exercise considerable influence on the economy as a result
of the purchasing power which they have in respect of those products of
agriculture and industry on which the white minority in control of
political affairs depends directly for its existence and survival.

It is quite obvious that unless far-reaching reforms are introduced
and early steps taken to review the policy of apartheid South Africa will
find it increasingly difficult to contain the situation, especially in view of
the pressures now being exerted on Rhodesia by America, Britain and
other Western countries to accede to African demands for majority rule
in that country in order to avert a possible bloodbath resulting from an
all-out racial war. It is known that South Africa favours these moves,
perhaps as a means of buying time for itself. But it cannot both support
the idea of majority rule in Rhodesia and refuse to make concessions in
furtherance of the same principle within its own boundaries. Considering the long, troubled and uncertain history of race relations in
South Africa, the resolution of South Africa's present dilemma cannot
be an easy matter. But it is quite certain that however long it takes for
majority rule, that is, rule by Africans, to be established in the country,
the present political and social trends in Africa and the rest of the world
will exert increasing pressure on South Africa to modify and eventually
abandon its obnoxious racial policies and move towards the idea of
a multi-racial society. So far the South African government has
responded to the various challenges that now face it by making only a
few token concessions to the coloured and African populations, but
both internal and external pressures for much more extensive and
fundamental reforms are rapidly building up, and important changes

will have to be made by the government within the next few years if a serious racial holocaust is to be averted.

THE FORMER BRITISH PROTECTORATES OF SOUTHERN AFRICA

Located in and around South Africa are the three independent African republics of Botswana, Lesotho and Swaziland. Lesotho, formerly known as Basutoland, is completely surrounded by South Africa, while Botswana (formerly Bechuanaland) and Swaziland are only partially enclosed, the former being bordered along part of its boundary by Rhodesia and the latter by Mozambique.

Despite their location, these territories were able to escape conquest by the Boers during the nineteenth century and became protectorates under the British Crown. No doubt one reason why the Boers did not persist in the conquest of these territories was the fact that at the time none of them was considered to be of much economic significance and therefore particularly desirable in the way that the Transvaal, for example, was. Nevertheless, for a long time after the creation of the Union of South Africa their political future hung in the balance because South Africa insisted that it had a natural claim to them, especially since their economies were so fragile that their principal means of subsistence was by exporting large numbers of their adult male populations to South Africa to work in the mines and urban centres as migrant labourers. It is estimated, for example, that as much as between 50 to 60 per cent of Lesotho's male population are to be found at any one time in South Africa, while the corresponding figures for Botswana and Swaziland are respectively about 27 per cent and 25–30 per cent.

The position of the protectorates was made even more helpless by the fact that, except in the case of Swaziland which had an alternative outlet to the sea through Lourenço Marques in the Portuguese colony of Mozambique, their only effective outlets to the outside world lay through South African territory. Indeed, so close were their political and economic ties with South Africa that during the days when they were British protectorates the British High Commissioner responsible for their affairs was based in South Africa.

After the Act of Union, South Africa continued to demand the incorporation of the three protectorates into the Union, and for a long time Britain partially conceded this claim by agreeing that there was the possibility of eventual incorporation, except that this would have to be done *after consultation* with the protectorates themselves. But what the term *consultation* meant was never clearly spelt out. In her weaker moments Britain insisted that it meant nothing more than merely informing the protectorates before handing them over, while at other times, when Britain's resolution was stronger, the interpretation given was that the territories concerned could not be handed over without the consent of their inhabitants.

At one time so sensitive was Britain to South African opinion in the administration of the protectorates that when in 1948 Seretse Khama, then heir to the chieftaincy of the Bamangwato tribe in Bechuanaland (now Botswana), and now President of his country, married an English woman he was exiled by the British government in deference to South African opinion, which regarded the marriage as an affront to white people and a direct challenge to South Africa's racial policies.

Thanks, however, to developments in other parts of Black Africa and the mounting pace of colonial emancipation, British policy towards South Africa gradually stiffened and her attitude to the political aspirations of the protectorates became increasingly liberal, resulting in the granting of full political independence to Botswana on 30 September 1966, to Lesotho on 4 October 1966 and to Swaziland on 6 September 1968.

The granting of independence to the protectorates was largely an act of faith forced upon Britain by trends in the rest of Black Africa. It did not, however, dispose of the problem of acute poverty faced by all the three countries concerned, which, with the possible exception of Swaziland, were largely dependent for their economic survival on the income derived from the export of labour to South Africa. Swaziland, however, was known to have considerable mineral wealth in the form of asbestos, high-grade iron ore and coal, mostly controlled by South African financial interests; and it also has a sizeable resident white population who own about 44 per cent of the land (Fig. 24, p. 171). Although these factors tended to give South Africa a particularly strong stake in the country's future, they nevertheless enhanced its economic viability.

Since independence, however, Botswana, whose arid conditions tended to make it the least favoured of the three countries, has made quite unexpected economic discoveries of copper deposits at two locations. Thus the only one of the three former British protectorates whose economic base still remains totally inadequate is Lesotho, which has traditionally been the largest exporter of labour to South Africa and therefore the most dependent of the three on the goodwill of the South African government for its economic survival.

THE FUTURE OUTLOOK OF THE FORMER BRITISH PROTECTORATES

Although Botswana, Lesotho and Swaziland are now independent, they are still subject to a number of serious constraints in the exercise of their political sovereignty on account of their unfortunate geographical location in relation to South Africa and the fact that their economic survival is so heavily dependent on South Africa, which both controls their main trade outlets and provides vital employment openings for large numbers of their populations.

Ever since these three territories succeeded in securing British protection round about the turn of the nineteenth century in their bid to avert conquest and assimilation by the Boers, South Africa has never reconciled itself to their separate existence outside of its political control and has found it even harder to come to terms with their recently acquired status as independent states ruled by Africans. While it is most unlikely that South Africa will wish to risk interference with the territorial integrity of any of these countries in the present world political climate, there is no doubt that it is fully conscious of their geographical and economic vulnerability and will not scruple to apply the most stringent sanctions and pressures against them if it ever feels any of its own vital interests threatened as a result of the activities of any of these countries.

The three countries concerned are thus placed in an extremely difficult position and are obliged to exercise the utmost caution and prudence in their dealings with South Africa, on the one hand, and, on the other hand, with the other countries of independent Africa which oppose South Africa's racial policies, however strongly they may feel about these policies.

SOUTH AFRICA'S NORTHERN NEIGHBOURS

North of the Limpopo River, which marks the northern boundary of South Africa, are the three former British territories of Rhodesia, Zambia and Malawi. All these countries have had British connections since about 1890.

Rhodesia and Zambia (formerly known as Southern and Northern Rhodesia) owe their foundations to Cecil Rhodes, who moved into the area from South Africa with the aim of developing a mining industry under British control to rival the rich mining area of the Witwatersrand, which at the time was under Boer control, and also in order to prevent the Portuguese who had made earlier incursions into the area from acquiring the whole of this central part of Africa by closing inland from their coastal possessions in Angola and Mozambique.

Rhodes was able to prevail upon Lobengula, the king of the powerful Matabele people in the western part of present-day Rhodesia who had themselves moved into the country as conquerors from South Africa, to grant him mineral rights in his kingdom, thus enabling him to establish a base for his British South Africa Company. Nine months later he despatched a pioneer column consisting of 200 settlers and 500 police from Bechuanaland to take possession of Mashonaland in the eastern half of the country. In the same year the king of the Barotse in the central part of present-day Zambia asked for British protection against the Matabele, who had extended their raids across the Zambezi River, and granted a concession over mining rights to Rhodes's Company, which by then had received a charter from the British government and authority to act as its agent in the region. But the apparent peace established with the local tribes was soon broken by a rebellion by the Matabele in 1893 and again in 1896, joined this time by the Shona. These uprisings were quelled by the British in 1896 and led finally to the annexation of the whole of Southern Rhodesia in 1897.

Meanwhile, the British South Africa Company continued to administer Northern and Southern Rhodesia with the support of Orders in Council made by the British government in 1889. Barotseland, however, had a different status from the other kingdoms; here, power was left largely in the hands of the paramount chief, while the Company confined itself to the control of commercial rights.

In 1911 the name Northern Rhodesia was given to Barotseland, although the Company still continued to rule it with a certain degree of Crown control until 1924, when it was placed completely under the administration of the British government with the status of a protectorate.

Company rule in Southern Rhodesia was virtually brought to an end in 1918, when the Privy Council's Judicial Committee ruled that the land belonged to the Crown and not to the Company, which could not therefore legally sell land at a profit to shareholders as had been done previously. After a financial settlement between the Company and Britain had been agreed upon a referendum was held for white electors to decide whether to join the Union of South Africa or to seek responsible government from Britain. By a majority of 3 to 2 the electorate voted for self-government, and in 1923 Southern Rhodesia was granted the status of a self-governing colony under the British Crown, but with Britain retaining control of external affairs and all legislation directly affecting the African population.

In contrast with Northern and Southern Rhodesia, which were colonised primarily for commercial and strategic purposes, Nyasaland began essentially as a missionary foundation, the most active bodies, following David Livingstone's pioneering efforts in the middle 1850s at suppressing the Arab slave trade in the area, being the Universities Mission, the Free Church Mission, and the Church of Scotland Mission. In 1891, in order to forestall an attempt by the Portuguese to extend their influence into the territory, Britain proclaimed it a protectorate and the Crown assumed responsibility for its administration. Even so the British South Africa Company nearly brought the country under its rule in 1896, but for the strong opposition put up by the missionaries, who felt deeply mistrustful of Cecil Rhodes and his supporters. In 1904 responsibility for its administration was transferred from the Foreign Office to the Colonial Office and its status accordingly changed to that of a colony.

It is apparent from what has been said above that although these three countries are contiguous their history since European settlement began has followed different courses, partly because of their inherent geographical conditions and partly because of the manner in which they came under European control.

Southern Rhodesia, which comes closest to South Africa in respect of the relations between the white and black races, is so different from Zambia and Malawi because, like South Africa, it was actually conquered by the Europeans, who proceeded to destroy its tribal system and to dispossess the Africans of their lands. This made it necessary for special reserves to be created for them under a scheme confirmed in 1930 by the Land Apportionment Act, which divided the country into native areas, European areas, and unassigned, undetermined and forest areas. Under this division the Africans received 33 per cent of the land, the small European population 50 per cent, while the remaining 17 per cent was held in reserve as unallocated land.

In contrast, Zambia was not conquered but peacefully penetrated by the British South Africa Company through a series of treaties which left the chiefs and traditional rulers with their political powers almost untouched and generally in full control of their lands. Only 6 per cent of the land was constituted into Crown lands which could be leased to Europeans, while 34 per cent was constituted into native reserves, and the remaining 60 per cent into Trust land earmarked for future use by the Africans. European settlement in Zambia was therefore of relatively minor importance, although the country's rich copper and other mineral deposits offered a lucrative avenue for economic exploitation by Europeans and led to a small colony of white settlers mainly along the country's central highland axis.

The position was different again in Malawi, which at the time of the arrival of the British was being mercilessly ravaged by Arab slavers and inter-tribal warfare. Accordingly, although force played a part in the European penetration it was directed as a result of the strong missionary influence in the country more towards the suppression of slavery and of inter-tribal warfare than towards the acquisition of land. No reserves were set up; rather the land was divided between freehold, public and African trust land. In theory both Europeans and Africans could acquire land on the same terms, but African interests were given paramountcy in the trust lands. Trust land formed 87 per cent of the total area, while public land and land held as freehold formed respectively 8 per cent and 5 per cent.

179

Thus, while in Rhodesia the initial idea of the British settlers and the government was to develop two entirely separate areas for white and black, in Malawi the idea was to make European and African development complementary and in Zambia, closely associated.

In time, however, Rhodesia drifted rapidly away from its two northern neighbours in its racial policies and came dangerously close to conditions in South Africa. But as the country was still under British rule, the British government continued to entertain the hope that the basic rights of the Africans would be safeguarded and, as in South Africa at the time of the Act of Union, inserted certain entrenched clauses directed towards this end in the constitution granting internal self-government to the country in 1923.

Unfortunately, these safeguards proved totally ineffectual. Having attained self-government, the white settlers of Rhodesia proceeded to follow the example of South Africa by keeping politics virtually as a white monopoly, firstly by making it difficult for all but a small number of Africans to qualify for the vote, and secondly by creating two separate electoral rolls for Africans and Europeans and severely limiting the number of seats in the legislature open to Africans so as to ensure a European majority at all times. But even these token concessions to the Africans caused widespread fears among the settlers and by the early 1950s moves were afoot to introduce a policy of outright racial discrimination aimed at finally ending all effective African participation in the government of the country. As in South Africa, the only role which the Europeans considered appropriate for the Africans was that of providing cheap labour in the mines and industrial establishments and on European-owned farms; but even in these areas various measures were adopted to keep the Africans out of the really skilled and highly paid jobs.

Such were the differences between these three countries; and they were well-known in British official circles and among the Africans living in the area. But what really brought them into world prominence was the decision by the British government to bring them within a single federation in August 1953 in the face of strong African opposition in the three countries and elsewhere in the continent and of liberal opinion in Britain itself.

THE ABORTIVE CENTRAL AFRICAN FEDERATION

From quite an early stage in their connection with Britain it was felt that steps should be taken to bring Rhodesia and Zambia or, as they were then called, Southern Rhodesia and Northern Rhodesia closer together politically. The first definite proposals for amalgamation were made in 1915, but this was opposed by the Europeans of Northern

Rhodesia for fear of being swamped by the larger European population of Southern Rhodesia, while the Europeans of Southern Rhodesia were apprehensive about the idea for fear that amalgamation with a territory which at the time was little developed economically would impose an excessive burden on their economy.

In 1929 the British government appointed a Commission to look further into the question, this time bringing Malawi (then Nyasaland) into the scheme (HMSO 1929). At the time a similar scheme for political amalgamation was under discussion for the three territories of British East Africa, and it seemed logical to extend the idea to the three British territories immediately to the south. The scheme was also opposed by the Europeans of Southern Rhodesia who again considered that Northern Rhodesia and Nyasaland were too poor to form suitable partners with their more prosperous country. But another reason for their opposition was the fear that union with these two countries lying so close to East Africa might draw Southern Rhodesia into an East African political system. Britain at the time had enunciated the principle of the 'paramountcy of native interests' in Kenya, and the Europeans of Northern and Southern Rhodesia feared that this principle might by a process of geographical attraction be extended to them and thus undermine their own position. What they favoured was a union between the two Rhodesias, but not one embracing Nyasaland, whose racial policies as far as Africans were concerned seemed to them to be unduly liberal.

In 1936 these demands for the amalgamation of the two Rhodesias were renewed, this time by the European settlers, but were rejected by Britain on the grounds that little change had occurred in the circumstances of the countries concerned since the matter last came up for consideration. However, in 1937 the British government appointed a Royal Commission under the chairmanship of Lord Bledisloe to enquire into the possibility of some form of closer cooperation or association between the Rhodesias and Nyasaland (HMSO 1939).

The report of the Commission, which was published two years later, accepted that the three territories would in the course of time become increasingly interdependent, but it opposed federation on account of the constitutional status of the three countries. However, in order to keep the objective of ultimate amalgamation in view, the Commission recommended closer economic cooperation and the establishment of an Inter-territorial Council to coordinate existing government services between the three countries.

As a result of the Second World War this Council was not established until 1945, but then as a purely consultative body. It was able to arrange for the extension of a number of Southern Rhodesian services to Northern Rhodesia and Nyasaland, but its effectiveness was severely limited by the lack of administrative machinery and funds for the implementation of its recommendations.

All this time, however, the Europeans in the three countries, who had been mostly involved in the discussions leading to these arrangements, had different motives for supporting the idea of closer political association. The Europeans of Southern Rhodesia hoped that through the federation they might more quickly attain the status of a full self-governing dominion, while the Europeans of Northern Rhodesia hoped that by joining the federation they would be able to escape from the control of the Colonial Office, which seemed wedded to the idea of the 'paramountcy of native interests'. The Europeans of Nyasaland were divided, some opposing it for fear that it would only result in an easier flow of labour from Nyasaland into the neighbouring countries, especially Southern Rhodesia, while others feared that it would undermine the freedom and welfare of the African population.

African opposition, as was to be expected, was not very vocal; but such as it was, it was motivated by fear of European political dominance. Opposition was particularly strong among the Africans of Nyasaland, whose African Congress was then looking forward to the development of Nyasaland as a self-governing country under a predominantly African government.

After the Second World War new developments occurred which made Southern Rhodesia even more interested in the idea of federation with its northern neighbours. Although European settlement had greatly increased and the country was spending large sums on development, its economic foundations were rather weak and cheap labour was in very short supply. In contrast, Northern Rhodesia had developed into a rich country with a booming copper industry, and Southern Rhodesia felt that a political union between the two countries would offer it important economic advantages.

Towards the end of 1950 the British government agreed that the question of closer union in the region should be examined afresh; and at a conference held in London in 1951, while the idea of complete amalgamation involving the unqualified transfer of power to a central body was rejected, agreement was reached on the creation of a British Central African federation with the proviso that matters affecting the day-to-day life of the inhabitants should remain under the control of the respective territorial governments.

Eventually in 1953, after a prolonged period of discussion and examination, and despite the clear opposition of the more advanced and vocal Africans to the proposal, the British government approved the establishment of a federation to be known as the Federation of Rhodesia and Nyasaland with a very complicated constitution aimed at safeguarding the interests of both Europeans and Africans and giving the British government the ultimate power in regard to the control of land as far as Northern Rhodesia and Nyasaland were concerned with a view to giving the Africans special protection in the apportionment of land. But even before the federation formally came into being it became apparent that these safeguards were not enough to protect the interests of the Africans and that ultimately they would find themselves wholly at the mercy of the European settlers, who dominated the African Affairs Board specially created to protect African interests.

In setting up the federation Britain hoped to provide an object lesson to the world in racial partnership and it was hoped that the African Affairs Board would exercise a moderating influence on the racial policies of Southern Rhodesia, where African interests were in the greatest jeopardy.

But the constitution for the federation was no more than a compromise which satisfied neither the Europeans nor the Africans. Thus, from the outset it was doomed to failure, and African opposition to it grew in intensity after Ghana's achievement of independence in 1957 and the start of the decolonisation process in black Africa.

On economic grounds the arguments in favour of the federation seemed very strong, since the economies of the three countries involved in it were in many respects complementary to one another. Southern Rhodesia, which was the most highly industrialised member and the one with the largest European population, began as a gold-producing country in the early part of this century and later became a major producer of chrome and asbestos (Fig. 24, p. 171). But more recently it had developed into an important industrial and agricultural country. Unlike its two northern neighbours, it had valuable deposits of coal at Wankie from which the copper mines of Northern Rhodesia obtained their fuel supplies. It also had some iron ore which had given it a basis for an iron and steel industry at Que Que, as well as other manufacturing enterprises. What it lacked, however, was an adequate supply of cheap labour and markets for its expanding economy, all of which Nyasaland and Northern Rhodesia were in a position to provide.

Nyasaland on the other hand was a small and poor country with a relatively large and well-educated African population which had gained a reputation in Central and Southern Africa as a source of efficient and intelligent labour. It is estimated, for example, that in 1960 nearly 160,000 Nyasa males were working outside the country, including 113,000 in Southern Rhodesia, 17,000 in Northern Rhodesia and 28,000 in South Africa. Its limited highland area, where the small European population was concentrated, offered a basis for commercial agriculture on a modest scale, but it had no significant mineral resources and hardly any manufacturing industries.

In contrast, Northern Rhodesia had rich copper deposits – among the richest in the world – and a few other minerals like zinc and lead; but it had practically no local sources of fuel for industrial development and relied mainly on imports from Southern Rhodesia (Fig. 24, p. 171).

A federation of the three countries thus offered an ideal combination for rational economic development, more especially in the light of the plans which had by then reached an advanced stage for the construction of the Kariba gorge hydro-electric power project on the Zambezi River designed to serve Northern and Southern Rhodesia, whose common frontier ran along the Zambezi.

But strong though the economic arguments in favour of federation were, the possible human consequences of the merger raised many doubts in the minds of those who were genuinely concerned about the welfare of the African populations and the protection of their political rights, especially in Nyasaland and Northern Rhodesia where many of these rights were still relatively intact.

It was obvious that the country which stood to gain most from the federation was Southern Rhodesia, which needed markets for its growing industrial output as well as an assured labour supply from its two northern neighbours, especially Nyasaland. Besides, the possibility of harnessing electricity from the Zambezi River, which forms a common boundary between Southern and Northern Rhodesia, was a matter of crucial importance to Southern Rhodesia's economic development; but this could not be realised without the cooperation of its northern neighbour. But the advantages of federation were not all on one side; Nyasaland also stood to gain from the arrangement because of the considerable financial subsidies to its limited economic base to be derived from a formal political association with Southern Rhodesia. The country which most resented the federation and was in a sufficiently strong economic position to do so was Northern Rhodesia, although the Advisory Commission on the Review of the Constitution of Rhodesia and Nyasaland which reported to the British government in 1960 indicated that there was substantial opposition to the federation from the white population of Southern Rhodesia, who feared that the inclusion of the two more liberally administered countries to the north within the federation would 'bring about a too rapid increase in the political power of Africans both in the Federation and in all three Territories' (HMSO 1960).

The constitution of the federation had provided for a review within not less than seven nor more than nine years from the date of its coming into force by a body made up of representatives from the federation itself, from each of the three constituent territories and from the United Kingdom chosen by their respective governments. The review commission was appointed in 1960, and the opponents of the federation took full advantage of it to voice out their sentiments and in the end left the members of the commission in no doubt as to the shortcomings and general impracticability of the federal arrangement. While the future of the federation was in the balance, a national uprising which had started in Nyasaland at the beginning of 1959 led to a concession by Britain in 1962 that that country could secede from the federation and, a year later, to a further decision that any of the constituent countries of the federation could secede if they chose. In June 1963 the representatives of the three territorial governments, together with those of the federal government and the United Kingdom, met at Victoria Falls to consider the orderly dissolution of the federation, thus bringing about at long last the ending of a political experiment which right from the outset had attracted hostility not only from the Africans in the three countries concerned, but also from many African countries as well as from large numbers of liberal-minded persons in Britain itself who had never been convinced of Southern Rhodesia's willingness to abandon or modify its oppressive racial policies or to respect the political rights of the Africans in the other two countries of the federation, contrary to the promises made to them at the outset. In December 1963 the federation was accordingly dissolved and Nyasaland and Northern Rhodesia decided to go their separate ways; and in July and October 1964, respectively, after the necessary constitutional agreements had been reached with London, the two countries were granted independence by Britain under the names of Malawi and Zambia, while Southern Rhodesia, now known simply as Rhodesia, reverted to its former status as a self-governing country under Britain.

Fortunately, the transition to independence in both Zambia and Malawi proved to be unexpectedly smooth. Zambia had never had a very large white settler population or one that was particularly attached to the land. As compared with the country's African population of some 3.4 million in the early 1960s, the European population numbered only about 75,000, while Asians numbered 8,000, and others 2,000. The European population, consisting mostly of miners, were concentrated around the country's railway along the central highland axis, with small outliers in the Fife–Abercorn (now Mbala) highlands and the Fort Jameson (now Chipata) areas in the north-east. The copperbelt extending from Lusaka, the capital, through Broken Hill as far northwards as the Zaire border was the real centre of European interest and the country's principal source of revenue. Thus, Zambia's entire political and economic future hinged on the extent to which the mining companies were prepared to cooperate with the newly established African government and on the general attitude of the tobacco-growing white settlers elsewhere in the country towards the changed political situation.

As it turned out, the mining companies gave the new African government their full cooperation even before it took office and thus in turn secured the goodwill of the country's new rulers led by Kenneth Kaunda. They had good reason to do so, since under the federation a large proportion of the country's revenues had been channelled into the central funds of the federation for the assistance of the other partners without any apparent corresponding benefits to itself.

Despite this promising start, Zambia's economy has been hampered since independence by its overdependence on copper, while the country's deteriorating relations with Rhodesia through which most of its copper and other exports find an outlet to world markets, as is shown later in this chapter, have created additional difficulties and turned it into one of the front-line states in the difficult diplomatic and armed struggle against white rule in Rhodesia and its South African allies.

Compared with Zambia, Malawi has had a relatively peaceful existence since independence as far as its relations with the white governments to the south are concerned, largely because of the country's consistent policy of avoiding active involvement in the fight against white domination in Rhodesia and South Africa. Malawi has instead striven for some type of political accommodation with the white dominated countries of South Africa in order to protect and develop its own economic well-being. This policy has been maintained in the face of strong opposition from other African countries.

As far as the white settler problem was concerned, Malawi was much more fortunate than its other neighbours. Out of the country's population of about 4 million in 1960 only 7,000 were Europeans, while Asians numbered about 10,000. The Europeans were mostly farmers, growing tea and tobacco in the Shire and Mlanje highlands. But the total size of their land holdings was small and as a group their attitude towards the question of African rights was considerably more liberal than that of their counterparts in the rest of the federation. But Malawi's economic base, as already mentioned, is extremely limited; it possesses hardly any mineral wealth and, although the alluvial soils along the shores of Lake Malawi and in the valley of the Shire River are quite fertile, their extent is so limited in relation to the high population of over 4 million that large numbers of Malawians are obliged to seek work outside the country. Added to these difficulties is the fact that the country is landlocked and is therefore dependent on its two neighbours, Mozambique and Tanzania, especially the former, for outlets to the sea and to world markets. Accordingly, Malawi's main problem since attaining independence has been how to develop a viable economy and raise the living standards of its people. The economic difficulties that face the country and its dependence on outside labour markets for large numbers of its adult male population are largely responsible for the inability of its government to adopt a firm stand against Rhodesia and South Africa which have been for many years, and still remain, the principal outlets for its surplus labour.

THE SPECIAL PROBLEM OF RHODESIA

With the dissolution of the Central African Federation and the granting of independence to Zambia and Malawi under black governments, the hopes of the white settlers in Rhodesia for the early attainment of independence suffered a serious setback, and the country was faced once again with many of the economic problems which had previously hampered its development. Although Britain had made a number of important concessions to the Europeans at the expense of African rights in order to win their cooperation as partners within the federation, it became increasingly clear that it would not be prepared to grant the country full political independence without a clear assurance that the basic rights of the African population would be fully respected.

As far back as 1960 Britain's attitude to the future of white rule in Africa had been clearly spelt out by the British Prime Minister, Mr Harold Macmillan, in his 'wind of change' speech delivered in Cape Town, in which he pointedly drew attention to the new political forces which were sweeping the countries of Black Africa irresistibly towards independence and Britain's general endorsement of these trends. Following the dissolution of the federation, many of the special safeguards which were entrenched in the Rhodesian constitution of 1923 for the protection of African interests were dropped by Britain, and it looked as if the white settlers of Rhodesia would have no difficulty in persuading the British government to hand over power to them in complete disregard of the possible fate of the African population. But this proved to be a serious miscalculation; Britain continued to insist on the principle of unimpeded progress towards majority rule and equality of treatment for all races as a condition for granting independence to the country. While Britain had no intention of handing over power immediately to an African government, it was quite obvious that, whatever its intentions might have been in the days when the empire was at its height, it had no intention now of allowing Rhodesia to achieve independence as a white man's country.

Rather than accept the conditions implicit in Britain's new stand on the country's political future, the European settlers decided to take over power by force; and on 11 November 1965 a new government elected almost exclusively by white voters unilaterally declared Rhodesia an independent country in complete defiance of Britain. The reaction of the independent states of Africa to this unexpected move was immediate. Since Britain still claimed to be legally responsible for Rhodesia, they demanded that Britain should intervene by force and protect the rights of the Africans in the country by arranging for an orderly transition to independence under a democratically elected government. However, although Britain accepted the need for ending Rhodesia's rebellion, it emphatically rejected the use of force as a means for dealing with the problem and sought instead the support of the United Nations for the application of economic sanctions against the country by all member states in the hope that the pressures and privations created by this measure would quickly bring the rebellion to an end.

The idea behind this strategy was that since Rhodesia is a wholly landlocked country dependent mainly on the port and rail terminus of Beira in what was then the neighbouring Portuguese colony of Mozambique for its trade outlets to the rest of the world the stoppage of the flow of Rhodesian exports and imports through Beira by means of a well-organised blockade would effectively strangle the country's economy within a relatively short space of time. Unfortunately, although this strategy produced some results, it proved less effective than had been expected, owing to a lack of cooperation from a number of Western countries who found other means of carrying on a clandestine trade with Rhodesia, while South Africa, as was to be expected, openly defied the embargo right from the start by offering Rhodesia an alternative rail outlet through its own ports, despite the long distances involved.

But although the sanctions failed to produce any immediate or dramatic results, it is quite obvious after ten years of operation that they have been sufficiently effective to induce the rebel government in Rhodesia to make overtures to Britain for a negotiated settlement over the question of independence. In addition, the mounting tempo of guerrilla activities organised by the African nationalist movement in the country with the assistance of the liberation movement of the Organisation of African Unity from bases in some of the countries bordering Rhodesia, especially Zambia and Mozambique, is beginning to convince the Rhodesian government of the need to pay more serious attention to the demands of the Africans for an equitable share in the government of the country.

Naturally, it is not easy for the white settlers, who number only 243,000 out of the country's total population of 5,400,000, to accept this proposition because it is obvious that any government elected on truly democratic lines will be dominated by Africans. To make the position of the settlers even worse, the former Portuguese colony of Mozambique, on which the country depends for its main trade outlets, achieved full independence under an African government on 25 June 1975 and has pledged its support to the cause of the African nationalists in Rhodesia, which means that the use of Beira and Lourenço Marques (now renamed Maputo) as outlets for Rhodesia's trade can no longer be taken for granted. In the meantime South Africa, has started making overtures to the independent states of Black Africa for peaceful co-existence and has begun to exert pressure on Rhodesia to adopt a more conciliatory attitude towards its black population. Obviously, neither the white settlers of Rhodesia nor the South African government genuinely want to see a black government in Rhodesia; but in the light of all the developments currently taking place, African rule in Rhodesia is now only a question of time.

The expected changes in Rhodesia (or Zimbabwe, as the African nationalists now refer to the country) could come even sooner than many people considered possible. Towards the end of 1976 follow-ing protracted negotiations, the heads of the five front-line states of Angola, Botswana, Mozambique, Tanzania and Zambia and the leaders of the African nationalist movements of Rhodesia, on the one hand, and the British and American governments backed by South Africa, on the other, agreed on a plan for the establishment of majority rule in Rhodesia by stages within a period of two years. In return for agreeing to this change the Western governments responsible for this initiative offered to provide Rhodesia with the capital needed to resus-citate its tottering economy and work for the end of the sanctions imposed against the country by the United Nations as well as of the guerrilla warfare waged against it by the African nationalist move-ments. They also agreed to set up a fund for helping and compensating any white Rhodesians who might lose their property or decide to leave the country as a result of the expected changeover to African majority rule.

It is too early yet to foretell the outcome of these proposals, which the white Rhodesian government openly admitted at the time that it was accepting with some reluctance. Certainly, the negotiations for putting them into effect have taken an unusually protracted course, and a number of people are now beginning to feel sceptical about their chances of success.

Right from the outset it was not at all clear how the proposals would be received by the African nationalists, especially the more militant factions, since a number of vital details regarding their actual implementation were never properly spelled out, thus making it poss-ible for each side to put its own interpretation upon them. So far the white minority government has done nothing to reassure the Africans of the genuineness of their desire to concede the principle of majority rule and hand over power to the African majority, and many of the African nationalist leaders are now convinced that the only way to bring about majority rule in the country is through armed struggle and not by negotiation with the minority white government. However, the initia-tives so far taken, however uncertain their outcome might appear to be,

represent at least an important first step in the long-outstanding process of decolonisation in the two remaining strongholds of white domination in Africa, for there can be little doubt that once majority rule has
been achieved in Rhodesia there will be increased pressure on South
Africa to make positive moves in the same direction.

ZAMBIA'S SEARCH FOR OUTLETS TO THE SEA

Among the many problems which the Rhodesian crisis has served to
underline in Southern Africa, one of the most serious without any doubt
is the dilemma which faces those states which lack direct access to the
sea. As has been shown, Rhodesia itself has been very hard hit by this
problem; but Zambia has suffered even more because of its location
right in the heart of the sub-continent.

Because of Zambia's main export of copper, which is a bulky commodity, the need for a ready and reliable rail link with the sea is
absolutely vital to its economic survival. At the time of Rhodesia's
unilateral declaration of independence (1965) there were three outlets
to the sea: one by rail through Katanga (now Shaba) in Zaire to the
Angolan port of Lobito; a second through Katanga and northwards to
the river port of Port Francqui by rail, thence by river transport to
Kinshasa and from there once again by rail to the port of Matadi which,
like Lobito, is on the Atlantic. This is obviously a difficult and expensive route and therefore not very practical. The third and most convenient route was that by rail across Rhodesia and Mozambique to the
port of Beira on the Indian Ocean. Before cheap electrical power
became available from the Kariba dam this route was particularly
convenient because it lay through the Wankie coalfield in Rhodesia,
thus making it possible for the rail wagons to load up with coal for
Zambia's copper industry on the return journey. However, following
the very strained relations which Rhodesia's illegal declaration of
independence produced between the two countries, the Beira route
which had served for decades as Zambia's main outlet for copper as
well as the principal artery for most of its imports had to be abandoned
and other alternatives hastily substituted.

In the initial stages this led to many difficulties and a special airlift
had to be organised with the help of Britain to keep the country
supplied with its essential fuel requirements in order to keep the
economy going. Thereafter, about 30,000 tons of the country's copper
output of over 600,000 tons per annum had to be sent by road to the
Tanzanian port of Dar es Salaam, which also handled most of its
essential imports, while the rest was evacuated by rail through Lobito.
But this arrangement was far from satisfactory; and accordingly in
1969 Zambia and Tanzania signed an agreement with China for the

construction of the Tanzam railway to link Zambia and Tanzania with a thirty-year interest-free loan of £150 million provided by China (Fig. 21, p. 158).

Work commenced in 1970 and was completed in 1974, and in October 1975 the railway was formally inaugurated. The line, which joins the main Zambian railway at a place called Kapiri Mposhi about 193 km (120 miles) north of Lusaka, runs for 1,871 km (1,162 miles) across very difficult terrain, especially in the southern section where it traverses the Muchinga Mountains of Zambia and the mountainous south-western portion of Tanzania, and its maintenance will not be an easy matter. But its completion has at last brought an end to Zambia's long and desperate search for a satisfactory outlet for its copper exports. It has also helped to isolate Rhodesia even further from the rest of the continent, while greatly strengthening the links between Zambia and Tanzania.

ANGOLA AND MOZAMBIQUE

Until their achievement of independence in 1975 the Portuguese colonies of Angola and Mozambique had the distinction of being the oldest European possessions in Southern Africa. Portuguese settlement in Angola began in the coastal area around the estuary of the Congo (Zaire) River in 1482, the same year in which the Portuguese built Elmina Castle in present-day Ghana, while settlement in Mozambique began along the coast round about the end of the fifteenth century in the wake of Vasco da Gama's great voyage of 1497–99 which pioneered the sea route to India. In 1576 the Portuguese founded Luanda, farther south along the Angolan coast, and had, at the beginning of the sixteenth century, established a number of strongholds on Mozambique Island and at Sofala along the Mozambique coast as well as at various inland points, such as Sena and Tete.

In both Angola and Mozambique, especially the former, there was considerable local opposition to Portuguese expansion inland from the coast, and it was not until well into the nineteenth century or even later that the two territories were fully pacified and their boundaries properly fixed. Before then Portugal had tried to extend its influence right into the heart of the continent in the region now forming Rhodesia and Zambia; but this attempt was forestalled by Cecil Rhodes, who won the area for Britain. But Portugal was a weak power and in no position to press its claims, especially since Britain had been largely instrumental in gaining international recognition for its claims to Angola and Mozambique at the Berlin Conference, despite its rather shaky hold of these territories, until their boundaries were firmly fixed a few years later.

Nevertheless, a clear indication of Portugal's intention to expand its Southern African empire and link Mozambique with Angola is provided by the prominent tongue of territory which extends along Mozambique's western boundary far into the heart of the continent along the Zambezi River, thus creating a conspicuous salient between present-day Zambia and Rhodesia, just as the Germans around the time of the Berlin Conference in their attempt to gain access to the Zambezi and Rhodesia for their South West African colony created the twenty-five to fifty mile wide Caprivi strip running some 250 miles eastwards along the northern border of what is now Namibia (Harrison Church 1956: Ch. xxxi).

As in their West African possessions, most of the initial efforts of the Portuguese in these two territories were concentrated on the export of slaves to Brazil. It is believed that in Angola, which suffered the worst depredation, over a million slaves were shipped across the Atlantic in the seventeenth century and another 2 million before the final collapse of the slave trade in the middle of the nineteenth century.

It was not until after the abolition of the slave trade that the Portuguese began to turn their attention to the development of agriculture and other productive economic activities. But the development of the two territories, whose combined area of slightly over 2 million sq. km (nearly 800,000 sq. miles) is more than twenty times that of Portugal itself, proved to be no easy matter for Portugal, which lacked both the financial and technical resources for the task. Yet because of these very inadequacies Portugal found it impossible to contemplate the abandonment of these territories on account of their vital role as a prop to its own shaky economy. In this respect, Angola was particularly valuable because of its obvious economic potential and the prospects which its extensive plateau tracts with their attractive climatic conditions offered for European settlement.

Right from the start, Portugal's colonial philosophy had rejected the idea of eventual political autonomy for the colonial subjects. Consequently, while the Portuguese paid commendable attention to the basic material welfare of the people of Angola and Mozambique and permitted social contacts between Europeans and Africans to take their natural course, they did little to advance the political development of the African population or to foster their indigenous institutions, which they summarily dismissed as primitive and undesirable. Rather, they held down their African subjects by means of the most oppressive measures and freely exploited their labour for state as well as for private Portuguese enterprises.

Portugal's colonial philosophy was taken a step further in 1951, when the country's ruler, Dr Salazar, transformed the political status of the overseas territories from colonies to overseas provinces, although in practice the change made very little difference to the lives of the vast majority of the African population. Illiteracy continued to be rampant and forced and contract labour continued to be exacted under the most inhuman conditions. Nor did the announced change in the status of the colonial territories, contrary to Portugal's expectations, restrain the United Nations and other interested bodies like the Organisation of African Unity from prying into the true state of affairs in these territories, even though according to the new definition of their status they were now part of Portugal itself and therefore supposedly sovereign and in no need of emancipation. And most of what was unearthed was highly disturbing and argued strongly for the early termination of Portuguese rule in Africa.

Before the Second World War neither Angola nor Mozambique had as yet succeeded in making any impressive economic advances, although Angola because of its relatively temperate climate was beginning to attract increasing numbers of Portuguese colonists, who were concentrated in the larger coastal towns, especially Luanda, and in the agricultural settlements around Nova Lisboa on the Bie plateau. In 1940 the Portuguese population in the country was no more than 44,000; but by 1960 it had risen to 172,000 and less than ten years later to 400,000. Mozambique was less suited to white settlement because of its hot, humid climate and the very limited area of highland which it contains, although after 1950 the Portuguese initiated a scheme for the settlement of about 5,000 white colonists every year in the more favourable arable valleys and along the principal railway lines. However, both territories occupied a highly important position in the transportation network of the northern portion of Southern Africa on account of their coastal location and the unquestioned command which this gave them over the rail outlets of the territories bordering them on the interior. The Benguela railway of Angola terminating at Lobito on the Atlantic Ocean served as the main outlet for the rich Katanga (now Shaba) Province of Zaire, while the Zambian copper belt and the industrial and agricultural area of Rhodesia centred on Salisbury were served by the Indian Ocean port of Beira in Mozambique. In addition, Mozambique's southern port of Lourenço Marques served as the ocean outlet for Swaziland and the northern part of the Transvaal in South Africa as well as for Bulawayo and the southern portion of Rhodesia within the Limpopo basin.

Post-war economic developments in the landlocked countries of Southern Africa have further enhanced the importance of Angola and Mozambique as trade outlets within the region. But other important economic developments have occurred within the two countries themselves, especially Angola, that have helped to reduce their former overdependence on the revenues derived from their railways and port facilities, even though these revenues continue to be a valuable source of foreign exchange.

Before the war Angola's exports consisted mostly of coffee, cotton and sugarcane as well as diamonds. Following the war, minerals began to assume a more prominent role in the economy; and today the country's exports include iron ore, copper and petroleum, while the agricultural settlements established by the Portuguese in the plateau regions at very great cost have begùn at last to show positive results (Fig. 24, p. 171). By 1969 the European population in the country had grown to about 500,000, mostly concentrated in the large cities, which were fast becoming miniature replicas of city life in Portugal.

Of all the economic developments that have taken place in Angola since the war, far the most outstanding has been the discovery of petroleum. The principal deposits are found in the coastal section of Cabinda; but other deposits occur north of Luanda and at Porto Amboim on the Angolan coast (Fig. 24). Exports of crude oil from Angola began in 1956 in quite modest quantities. However, with the discovery of the Cabinda deposits in 1966 production began to soar dramatically, rising from 537,000 tons in 1967 to 752,000 tons in 1969, a year after the exploitation of the Cabinda field began. By 1971 production had reached 5,700,000 tons, making Angola the sixth largest producer in Africa. Most of the oil is exported in the crude state, but in recent years a newly built refinery at Luanda has begun to export some finished petroleum products.

These developments gave immense satisfaction to the Portuguese. They were seen as the most effective answer to Portugal's enemies and the clearest justification for its much-criticised colonial policies and practices. But in spite of all these material achievements the question of the political rights of the people of Angola and Mozambique and their future role in the direction of political affairs in their own countries still remained unresolved.

In 1961, however, Portugal's complacent attitude towards its African colonies received a rude jolt when the people of Angola began a massive rebellion against Portuguese rule, followed in 1963 and 1964, respectively, by similar rebellions in Guinea Bissau and Mozambique. Within a few years the rebellions had assumed the proportions of a full-scale war, which continued to rage until 1974, when its repercussions on Portugal's economy and on the morale of the Portuguese army culminated in the overthrow of the Portuguese government in Lisbon by a military coup. The new military government immediately announced its intention to negotiate for a peaceful settlement with the various nationalist movements whose armies were ranged against Portuguese troops in the African colonies and to grant independence to the local population.

No one had expected that the Portuguese territories in Africa would achieve independence in such a dramatic fashion or that the wars of liberation started by their various nationalist movements would have the effect of liberating Portugal itself from the authoritarian government which had ruled the country for so many decades.

On 25 June 1975 Mozambique achieved its independence, and Angola followed a few months later in November of the same year. As was to be expected, the transition to full political independence did not prove an easy matter, especially in Angola where no less than three rival nationalist movements based mainly on tribal affiliations were struggling for ascendancy; and the transfer of power to an African government was further complicated by the large number of Portuguese settlers in the country, most of whom had come to regard Angola as their home. Matters were further complicated by the fact that while preparations were going on for the granting of independence to Angola a dissident movement, again mainly tribal in origin, began in the oil-rich Cabinda province with the presumed backing of Zaire, while the rivalry between the nationalist movements in Angola itself – the Popular Movement for the Liberation of Angola (MPLA), the National Front for the Liberation of Angola (FNLA) and the National Union for the Total Independence of Angola (UNITA) – developed into a bloody civil war, resulting in the loss of many innocent lives and the wholesale evacuation of thousands of Portuguese settlers and of other European nationals from the country. Individual efforts by a number of African leaders as well as by the Organisation of African Unity to bring about a peaceful settlement proved of no avail. In the midst of the turmoil the Liberation Front of Cabinda, which for some time had been campaigning for secession, unilaterally declared Cabinda independent on 1 August 1975, thus further adding to the confusion and effectively robbing Angola of the vital oil resources of the territory, which were among the major economic prizes the Angolan nationalists expected to secure on the attainment of independence.

The full effects of these tragic developments on the future course of events in Angola still remain to be seen, but their seriousness cannot be doubted. The situation in the country immediately following the granting of independence was reminiscent of what happened in Zaire in 1960, when that country achieved its independence from Belgium without any serious preparation, after years of oppressive and paternalistic colonial rule.

Following the failure of the Organisation of African Unity to reconcile the opposing factions the civil war dragged on until in 1976 the MPLA clearly gained the upper hand and received general international recognition as the country's government. It will take some time before complete political calm and stability are restored in Angola, and even now (in 1977) sporadic fighting continues in various parts of the country. But Angola has now taken its place as an independent state within the Organisation of African Unity and it is unlikely that the MPLA government can be dislodged easily by any of its rivals even though the pro-Marxist orientation of the government is resented by the Western nations which formerly had vested economic interests in the country. What remains now is for the extensive damage caused by the civil war to be repaired as quickly as possible so that national unity may be restored and the economy placed on a sound footing. The problems facing the country are indeed enormous, for the war reduced economic output by half and led to the exodus of most of the white population who had held practically all the key positions in the vital sectors of the economy, while thousands of Angolans – soldiers as well as civilians – were killed. The task of national reconciliation is made all the more difficult by the fact that the various nationalist factions are divided along sharp ideological lines, with the MPLA generally enjoying the backing of the Soviet Union, while the FNLA and UNITA are pro-Western in their orientation. But what is certain beyond all doubt is that at long last this highly prized Portuguese colony is now fully under the control of an African government which is determined, like most of its black neighbours, to work relentlessly for the emancipation of the whole of the rest of Southern Africa from white domination.

In contrast with Angola, Mozambique appears to have succeeded in effecting a fairly smooth and peaceful transition from colonial status to independence and thus to be well set for making an effective contribution to the current efforts of the independent African states towards the complete emancipation of the rest of Southern Africa from European political domination. No doubt one important advantage which Mozambique has had over Angola is that its struggle for independence has been conducted mainly by a single nationalist movement – the Mozambique Liberation Front (FRELIMO) – which has now taken over the administration of the country from the Portuguese. Nevertheless, the country is faced with a number of serious problems which stem from its geographical location and the somewhat limited nature of its economic base.

For years Mozambique's economy has depended heavily on earnings derived from services rendered by its railways and ports to the countries which border it on the interior, especially Rhodesia, South Africa, Zambia and Malawi, as well as on cash remittances sent into the country by or on behalf of the thousands of Mozambique citizens who are employed as labourers in the mines and industrial establishments of Rhodesia and South Africa. By a long-standing agreement with the Portuguese government, South Africa is allowed to recruit up to 80,000 Africans annually from Mozambique for work in the Witwatersrand.

In return Mozambique receives taxes from the employers as well as a proportion of the wages of the migrants, amounting to about £1 million annually, which is paid to the workers in local currency on their return to Mozambique, although the Portuguese government is allowed to recover an equivalent amount in South African bullion. Rhodesia also has for many years provided regular employment for about 100,000 workers recruited from Mozambique every year.

The railways also bring in considerable revenues. The oldest line is the one linking the port of Lourenço Marques with the Transvaal border. It was built at the end of the nineteenth century and serves today as the outlet for much of the industrial and mining traffic of Johannesburg and the Rand. Under the agreement for the supply of workers from Mozambique to South Africa, Lourenço Marques is guaranteed 47.5 per cent of the transit trade from the gold mining region of the Transvaal. A second railway links the port of Beira to Umtali on the Rhodesian border, from where it continues to Salisbury and beyond. Rhodesia depended heavily on this line for its exports and imports and until Zambia's break with Rhodesia a few years ago, it also served as the main outlet for Zambia's copper exports. A third line runs from Beira and over a long bridge across the Zambezi and then on into Malawi, through Blantyre to Salima near Lake Malawi. This line is Malawi's main outlet to the sea. Finally, there is the railway, completed in 1954, which links Lourenço Marques to the south-eastern part of Rhodesia, where a considerable amount of agricultural expansion has taken place during the past few years. Altogether, Mozambique has over 3,542 km (2,200 miles) of railways and, until the closure of its boundary with Rhodesia, it was estimated that something like two-thirds of the rolling stock was used for international traffic. Now that Mozambique has achieved independence under an African government which is committed to the liberation struggle in the continent, it is quite obvious that all or most of these various trade and labour agreements entered into between the former Portuguese government of the country and the white regimes in Rhodesia and South Africa will have to be reviewed. The question is made particularly difficult by the fact that Mozambique itself stands to gain as much from the continuation of these arrangements as these two neighbouring countries, if not more so, because of its limited resource base and the relatively backward state of its economy following years of neglect by the Portuguese and the effects of the long war of liberation from colonial rule. Another complicating factor in any attempt by Mozambique to sever its communications links with the two countries, especially Rhodesia, is the fact that some of the communications involved also serve friendly African countries which lack direct outlets to the sea. Yet it is clear that by opening its ports and railways to Rhodesia and South Africa without any kind of restriction and by keeping their mines and industries supplied with cheap labour Mozambique would be advancing the economic and political interests of these countries in no small measure.

In the face of the difficult alternatives presented by these complex problems the Mozambique government has adopted what is plainly a compromise solution based on the country's own vital economic interests and the obligations imposed by its commitment to the African liberation struggle in Rhodesia and the rest of Southern Africa. So far it has been felt necessary only to change the country's economic policies towards Rhodesia. In March 1976 Mozambique closed its border with Rhodesia, thereby making a very important contribution to the effectiveness of the trade sanctions imposed some years ago by the United Nations against that country but at the same time forfeiting one of its own major sources of revenue. In compensation for this loss and also to help the country to re-establish its economy the United Nations, the Organisation of African Unity, the Commonwealth and various outside organisations and countries have provided Mozambique with generous financial and other forms of assistance. As far as South Africa is concerned, however, trade and labour relations remain much as they have always been, although understandably not much prominence has been given to this aspect of Mozambique's foreign policy. Indeed, there are even now plans to sell electricity to South Africa from the Cabora Bassa dam, thus further strengthening the economic ties between the two countries. At the moment about half of Mozambique's foreign exchange earnings are derived from South Africa.

While the ambiguities in Mozambique's policies towards Rhodesia and South Africa may appear strange to many people, especially in view of Mozambique's strong nationalist and anti-colonialist policies since its achievement of independence, it has to be recognised that, like so many other countries in this and other parts of the world, Mozambique is not an entirely free agent in determining its economic and political destinies. Thanks to the country's geographical location and its historical experience, it has been forced into certain economic relationships with these two white-controlled neighbours whose disentanglement, however necessary it may be, is bound to be costly, inconvenient and time-consuming, and any attempt to force the pace unduly on emotional or ideological grounds could do serious harm to Mozambique itself.

9
East Africa

East Africa is the smallest of the political regions of Africa, although in absolute terms its area of 1,763,769 sq. km (680,992 sq. miles) represents quite a sizeable stretch of territory. The region is made up of only three countries – Kenya, Uganda and Tanzania, including the two small islands of Zanzibar and Pemba (Fig. 25). On strictly geographical grounds Rwanda and Burundi could also be said to belong to it, but as has been shown in Chapter 7 the political and economic orientation of these two states, which once formed part of the German colony of Tanganyika out of which Tanzania was born, was directed towards Central Africa after the First World War, when they were transferred to Belgium as the mandated territory of Ruanda-Urundi and subsequently brought under a common administration with the Belgian Congo, just as the rest of Tanganyika was transferred to Britain.

From both the physical and the human standpoint, East Africa shares many similarities with Southern Africa; and indeed along the common frontier between the two regions it is difficult to determine with precision exactly where one begins and the other ends.

This is hardly surprising, since both regions belong to Highland Africa and the high plateau which is such a conspicuous feature of the physical geography of Southern Africa extends without a break into East Africa, except that in the latter region it attains even higher elevations and is characterised by even more spectacular scenery on account of the widespread fracturing and tilting which produced the rift valley system of East Africa together with its accompanying volcanoes and lava flows during the Tertiary era. As a result of these high altitudes and the climatic conditions produced by them both regions have been associated with permanent white settlement, thus further differentiating them from the rest of the continent and saddling them with a formidable array of social, economic and political problems, whose effects are still being felt even in those countries that have achieved full political independence under African governments.

For many years the presence of white settlers in East Africa and Southern Africa was the outstanding political characteristic of the two

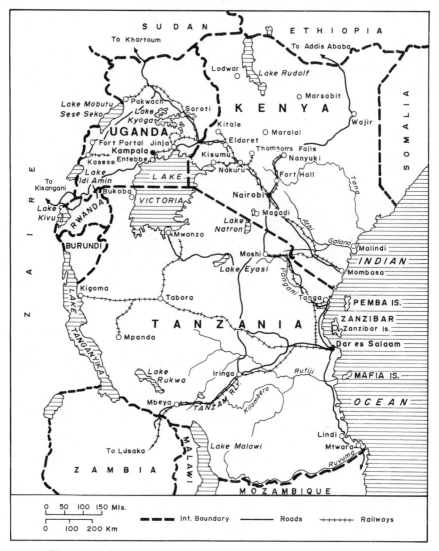

Fig. 25. East Africa: political divisions and communications

regions, even surpassing in importance the fact that the indigenous inhabitants of both of them consist predominantly of peoples of Bantu stock. Finally, by what can only be described as an accident of history, it so happened that the dominant colonial power in both regions was Britain, whose influence, following the elimination of Germany from the African colonial scene after the First World War, became firmly established over a continuous expanse of territory stretching from the Cape of Good Hope all the way along the highland spine of the

200

continent as far as the northern boundary of Kenya and Uganda and beyond that through Sudan and Egypt to the southern shores of the Mediterranean Sea.

The strategic significance of this impressive chain of British-held territories overlooking the Indian Ocean, where some of the most valuable possessions of the British Empire were located, was obvious and greatly strengthened Britain's determination to consolidate its hold of its three East African possessions which formed a vital link in the chain (see Fig. 8). As far as the African populations of East Africa and Southern Africa were concerned, however, the British presence gave them yet another bond of unity and a means of ready intercourse through the use of the English language and their common exposure to British political and social institutions.

In addition to altitude and climate, a factor of great importance in the political geography of East Africa has been the comparative sparsity of the population. In relation to its area, East Africa's population of 40 million is very small; but the disparity is seen to be even greater when the three countries of the region are considered separately, as shown in the table below. In Uganda, which is the smallest of the three, the population density is quite high, but Kenya and Tanzania are grossly underpopulated. Several reasons can be given for this state of affairs. Whereas in Uganda the climate and other environmental conditions have generally favoured agriculture and human settlement throughout the country, in Kenya and Tanzania large tracts of territory have been rendered quite unsuitable for human occupance by low or unreliable rainfall or by the prevalence of such serious environmental diseases as malaria and trypanosomiasis. In addition these two countries have suffered much more from the devastating effects of the Arab slave trade, which caused extensive depopulation in the region immediately before the arrival of the colonial powers. Even during the colonial era Tanzania continued to suffer from serious depopulation on account of various epidemics and the large-scale destruction of life which resulted from the German pacification of the country.

Political divisions	Area		Population 1975 or latest
	Sq. km	Sq. miles	
Kenya	582,646	224,960	13,399,000
Tanzania	945,087	364,898	15,155,000
Uganda	236,036	91,134	11,549,368
Total	1,763,769	680,992	40,103,368

There is little doubt that the relative emptiness of East Africa at the time of the European arrival in the region and the abject and demoralised state to which the local population had been reduced by the Arab slave trade helped greatly to facilitate the establishment of European rule in the region, especially in Kenya and Tanzania which, unlike Uganda, appeared to be without any well-organised indigenous kingdoms capable of resisting outside intrusion.

EUROPEAN COLONISATION AND SETTLEMENT

As in other parts of Africa south of the Sahara the pioneers of European colonisation in East Africa were the Portuguese, whose contact with the region began with Vasco da Gama's visit to the coast in 1498. A century later, however, they were driven out of the area by the Arabs. For the period of nearly three centuries which followed, Arab traders and slavers operating from bases along the coast assumed control of the region and carried on a ruthless and devastating traffic in human beings who were mostly exported to the Middle East.

It was not until the second half of the nineteenth century that European interest in the region was revived, mainly through the enterprise of British explorers, missionaries and philanthropists, whose initial aim was the suppression of the slave trade, the conversion of the African population to Christianity and the development of more normal forms of trade and commerce. Inevitably, the activities of the missionaries and traders led to their assumption of various obligations to the African population, which in due course proved too much for the limited resources of private organisations and individuals and thus led to the intervention of the British government.

The first part of East Africa to come under British influence was the coastal portion of Kenya, where the Sultan of Zanzibar had long been in control. In an attempt to gain a footing for the abolition of slavery Britain appointed a consul at Zanzibar in 1840. Shortly before then the Germans had begun to extend their political influence in the region, thanks to the activities of the enterprising German explorer, Karl Peters. However, their progress was delayed by the revolt of the Arabs in the second half of the nineteenth century. Although the revolt was suppressed in 1889, it nevertheless occupied the attention of the German administration and slowed down their advance inland.

British explorers, notably Speke and Baker, had reached the area of Uganda in the 1860s in their efforts to discover the sources of the Nile. They were so impressed by the natural wealth of the area and the advanced social organisation and institutions of the people that in 1888 a British East Africa Company was formed with the object of developing the economic resources of the area. The Company successfully contested the claim of Karl Peters to Uganda until the British government took over responsibility for the area in 1893. The Company similarly looked after the interests of Kenya until its charter was surrendered to the British government in 1895. The area was declared a British protectorate and the construction of a railway from Mombasa to Uganda began at the end of the century. Meanwhile, in 1890, Zanzibar and the adjoining coast of Kenya had been turned into a British protectorate.

The decision of the British government to construct the railway from Mombasa to Uganda marked an important turning point in the history of East Africa by establishing a permanent British presence in the region.

The real pioneer of white settlement in Kenya was Lord Delamere, who visited the country in 1897. He saw the possibility of reproducing European agriculture in the country and opened the eyes of the British government to the prospects which this offered for paying for the high cost of the Mombasa–Uganda railway, and accordingly the British in 1902 adopted in embryo a deliberate policy of European settlement in the area (Fig. 26).

East Africa has many contrasting landscapes, but from the point of view of white settlement the most attractive part was the central portion of Kenya with its lofty volcanic mountains and its cool climate and rich soils. Thus, while Uganda and Zanzibar remained as protectorates where local institutions and land rights were left largely intact, Kenya quickly became a British colony in the full sense and vigorous efforts were made to promote white settlement and white agriculture there. For climatic reasons white settlement was out of the question in the low lying equatorial island of Zanzibar. In Uganda both climatic and soil conditions were favourable, but white settlement was not an easy proposition because of the advanced and relatively densely populated societies which the British encountered, especially in the old kingdom of Buganda whose advanced culture and Byzantine-like court procedures had so impressed Speke when he first visited the royal court of the Kabaka in 1862.

In Tanganyika, also, the Germans found an attractive area of highland around Mount Kilimanjaro and the Usambara Mountains, but these were not as extensive as the highlands of Kenya. In the southwestern part of the country there were other areas of highland, notably the Livingstone and Rungwe Mountains, but again these were limited in extent and had the additional disadvantage of being isolated and widely scattered. Although the Germans made a start at white settlement, their efforts were cut short by the First World War, and it was left to Britain as the mandatory power to continue the process, although in this case this had necessarily to be done under the scrutiny of the League of Nations (Fig. 26).

At the same time that white settlers were being encouraged to come to East Africa by Britain the immigration of Indians into the area was

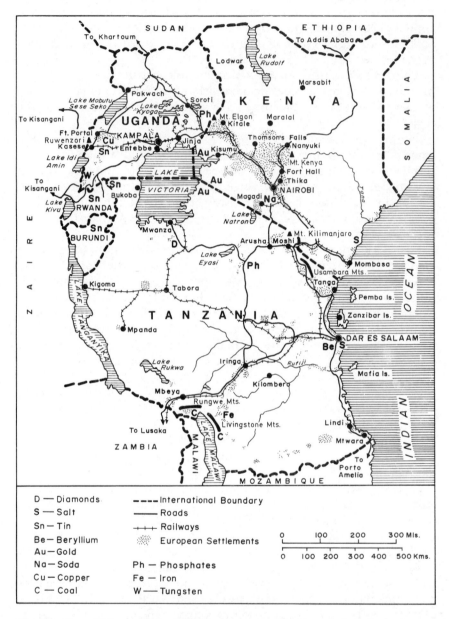

Fig. 26. East Africa: political divisions, communications, minerals and areas
of white settlement

also proceeding. From the earliest times Indian traders had been operating along the coast, and as the hinterland was opened up many of them moved inland. But it was the demand for labour for the construction of the Mombasa–Uganda railway that really attracted large numbers of Indians into the interior. However, attempts were also made to encourage the settlement of Indian peasants on the land as a deliberate policy, especially in the parts of the country just outside the white areas, although these were not particularly successful. Nevertheless, the presence of the Indians soon began to be seen as a threat to the white settlers, and thus led to the formulation of a White Highlands Policy aimed at this stage primarily at keeping out the Indians rather than the Africans from the white areas.

The area which the Europeans favoured was the Kenya highlands (including the portion of the rift valley located in the area) which they conveniently assumed to be available for white occupation. What they did not realise or, more probably, refused to recognise was that the real reason for the apparent emptiness of the highlands was that the Kikuyu, to whom the area belonged, and who were a predominantly agricultural people, had been kept out of these lands by the raids of the semi-nomadic Masai herdsmen. However, at the time the British arrived the Masai were no longer such a serious menace because of two severe epidemics which had hit their cattle in 1884 and 1889 and another devastating epidemic of smallpox which had afflicted the tribe itself and reduced them to a condition of virtual impotence as warriors.

Thus, without considering the possible claims of the local population to the highlands, the British proceeded to acquire land in the area on an extensive scale, and in 1901 the East Africa Lands Order in Council was made to give a legal backing to the acquisitions of the settlers, thus bringing into being the White Highlands of Kenya. However, in order to alleviate hardship to the dispossessed African population and also to make available a conveniently located reservoir of labour for the settlers it was decided in 1904 to create a number of African reserves mostly on less attractive land within and around the White Highlands. Some of these reserves were marked by intense overcrowding with the consequent rapid deterioration of their poor soils. In contrast, the best lands forming the greater part of the highland zone of some 41,474 sq. km (16,000 sq. miles) were assigned to the Europeans, thus causing widespread dissatisfaction among the Africans, especially the Kikuyu.

Added to the problems created by this inequitable distribution of land as between Africans and Europeans, the British government feared that demands for land in the highlands would also be made by the large and growing number of Indians, many of whom had been brought into the country at the beginning of the century as labourers for the construction of the East African railway.

To settle all further doubts about the rights of the Africans in the country, and in particular to avert any possibility of Indian claims to a share of the highlands the British government issued in 1923 the famous Devonshire declaration regarding the 'paramountcy of native interests' which, among other things, stated as follows:

Primarily, Kenya is an African territory, and His Majesty's Government think it is necessary definitely to record their considered opinion that the interests of the African natives must be paramount, and that if, and when those interests and the interests of the immigrant races should conflict, the former should prevail. Obviously the interests of the other communities, European, Indian and Arab, must severally be safeguarded. Whatever the circumstances in which members of these communities have entered Kenya, there will be no drastic action or reversal of measures already introduced, such as may have been contemplated in some quarters, the result of which might destroy or impair the existing interests of those who have already settled in Kenya.

But in the administration of Kenya His Majesty's Government regard themselves as exercising a trust on behalf of the African population, and they are unable to delegate or share this trust, the object of which may be defined as the protection or advancement of the native races [HMSO 1923].

As far as the exclusion of Indians from the White Highlands was concerned, the Devonshire declaration achieved the desired aim, but despite its lofty sentiments it failed disastrously to safeguard the proper interests of the Africans both as regards the distribution of land in the highlands and their political rights in the general administration of the country.

The land question continued to agitate the minds of the Africans, and in order to put an end to their nervousness the British government decided to appoint a Land Commission in 1932 to look into the whole matter. The findings of the Commission brought little comfort to the Africans; after defining the White Highlands as comprising an area of 43,936 sq. km (16,950 sq. miles) the Commission went on to recommend that no Africans or non-Europeans should be permitted to hold land in the area and that any such persons already there should be removed. This dealt the final death-blow to the hopes which the Africans had long entertained for a fair deal in the allocation of land, while giving new hope to the European settlers who had seen in the Devonshire declaration a serious threat to their future security. At long last white settlement in Kenya became firmly established with the backing of the British government despite the widespread discontent which it generated among the African population.

By contrast with developments in Kenya, European colonisation in Uganda followed a relatively smooth and peaceful course. After Speke's famous visit to the country in 1862 the next European contact came with Baker in 1869, who as governor-general of the Egyptian province of Equatoria tried to destroy slavery and to control Uganda from the north. But he achieved little success owing to the daring activities of the Egyptians and the Turks in the kingdom of Bunyoro which until the nineteenth century had been the most pre-eminent of the local kingdoms. For years Bunyoro had resisted European penetration and served as a refuge for dissident factions from the more peaceful kingdom of Buganda and for Arab and other slave traders.

Much of the subsequent history of Uganda became bound up with missionary activity, both British and French missionaries forming foci for political rivalry between their respective countries for the control of the area. In 1894, however, as a result of the work of British administrators such as Captain (afterward Lord) Lugard a British protectorate was proclaimed over Uganda, following a brief period of rule by the British East Africa Company. Really effective administration came with the regime of Sir Harry Johnston as commissioner of the protectorate in 1900, but even so the old rivalries continued among the principal local kingdoms of Bunyoro, Buganda, Ankole and Toro (Fig. 27). However, by 1914 the whole country was effectively under British administration and its boundaries were fixed.

Despite Uganda's attractive climate and the obvious prospects it held for profitable agriculture, the British administrators came to the conclusion early in the twentieth century that the country was not suitable for white settlement, no doubt, as already pointed out, on account of the comparatively high density of the African population and their advanced social and cultural development. This view was upheld by the British government and adopted as official British policy. Instead of attempting to establish white settlements, therefore, the British turned their attention rather to the encouragement of African commercial agriculture as a means of defraying – at least in part – the enormous cost of the Mombasa–Uganda railway, and this led to the promotion of cotton cultivation, later to be reinforced by coffee. Unlike Kenya, where at a certain stage Africans were prohibited from growing certain crops, especially coffee, in Uganda most of the commercial agriculture was carried on by Africans, and the few European settlers in the country played a relatively minor role. Here, also, there was a considerable number of Indians who had come in with the railway, but their main area of activity was commerce rather than agriculture, and they therefore posed no problem as far as the allocation of land was concerned.

As far as the apportionment of land was concerned, the British government pursued fairly liberal policies under various agreements reached with the rulers of Buganda, Ankole and Toro. By the Uganda Agreement of 1920 freehold estates known as *mailo lands* amounting to approximately 50 per cent of all the available land were set aside for the

Fig. 27. The major African kingdoms in Uganda at the turn of the nineteenth century

use of the chiefs and people of Buganda and the rest treated as Crown land which the government was free to allocate in its discretion, while in Ankole and Toro the *mailo lands* amounted to only 6 per cent of the available land. In contrast, Bunyoro was treated as conquered territory and all lands were accordingly placed under the control of the Crown until 1933, when under new agreements a new system of land allocation was introduced.

It was from the Crown lands that the comparatively small number of European settlers in the country obtained allocations for their estates; but the total area thus granted to them was quite small by comparison with the area set aside for exclusive European occupation in Kenya.

As we have already noted, Tanganyika (now Tanzania) began as a German territory. Here also the activities of missionaries helped to pave the way for European rule, but it was the German explorer, Karl Peters, who really won the territory for Germany. By agreement with the Sultan of Zanzibar the Germans established a sphere of influence in the interior, which was administered for a brief period by the German East Africa Company, while the Sultan retained control of the ten-mile wide strip along the coast. Under an agreement reached in 1886 with the British, the respective spheres of the two powers were defined and in 1890, under another agreement, their boundaries were delimited. Meanwhile, the German East Africa Company had succeeded in obtaining a lease of the coastal strip from the Sultan in 1888, but their tactlessness led to a local rising by the Muslim population, which was only put down after the intervention of the German government. In the light of this experience it was decided to end company rule and in 1891 Germany declared a protectorate over the entire territory. The actual consolidation of the territory lasted until 1906 and was marked by a series of wars, punitive expeditions and suppressive campaigns which resulted in the large-scale destruction of human life over extensive areas, especially in the south during the Maji-Maji rebellion of 1905.

Right from the outset the encouragement of European settlement formed an integral part of German policy in Tanganyika or German East Africa, as the country was initially called. By 1917 the government had transferred to immigrant – largely European – ownership about 526,082 ha (1,300,000 acres) of land, mainly on the coast and in the northern highlands around Mount Kilimanjaro.

After the First World War, however, this programme of white settlement suffered a marked decline owing to the liquidation of German property in the country and general world economic depression of the 1930s. Besides, since Britain was now in control of the country as the mandatory power, its policies in Tanganyika could not but be influenced by the principle of the 'paramountcy of African interests' which it had enunciated in Kenya in 1923. It was accepted under the mandate that Tanganyika was to be developed for the benefit of the African population, and European settlement was therefore encouraged only where it was judged to be in the interest of the Africans. As for the large number of Indians who had found their way into the country over the years, they functioned mostly as a commercial class of small traders, middlemen and artisans.

A major obstacle to the development of the country in the early stages was its large size and the sparsity of the population. The Germans made a start with railway construction by building a line from the port of Dar es Salaam to Kigoma on Lake Tanganyika, which was later linked by the British to Mwanza on Lake Victoria as well as another shorter line in the north from Tanga to Moshi, both of which were completed by 1914. These railways helped greatly in the promotion of commercial agriculture; and by the time the Germans left a sound foundation had been laid for the cultivation of cotton and coffee, mostly by African farmers, and of sisal by Europeans.

The modern economic and political development of the country, however, really dates from the end of the First World War with the changeover to British administration. That was when the name Tanganyika was adopted. Under the governorship of Sir Donald Cameron, which lasted from 1925·to 1931, important advances were made in several fields, especially in local administration, which was based on the then novel philosophy of indirect rule. Another notable development was the establishment of a Legislative Council composed of British officials as well as European and Indian unofficial members, a development which placed Tanganyika ahead of Kenya, where the European settlers up to that time had no representation in the govern- ent. Cameron's predecessor in office had enacted a Land Ordinance in 1923, which ensured the security of African land rights. With this important enactment as a basis, the reforms introduced by Cameron set the country on a course of economic and political development very similar in essentials to trends in Britain's more progressive West African colonies, except for the absence of African representation on the country's Legislative Council. Today, the leading exports are sisal, coffee and cotton, together with diamonds which were first discovered in commercial quantities in 1940.

EMERGENCE OF INDEPENDENT STATES

Really serious efforts towards the achievement of political independence in East Africa date from after the Second World War, but the pace varied considerably in the three countries of the region, depending on local circumstances. Because of the presence of large numbers of well-organised and articulate white settlers in Kenya the political future of the Africans there seemed much less certain than in Uganda and Tanganyika.

Up to 1944 only Europeans participated in the government of Kenya. In 1944 the first African member was appointed to the Legislative Council, an event which was hailed in British official circles as heralding a new principle of partnership between the different races, not only in Kenya but in the whole of British East Africa. The term 'partnership' which had first been proposed as far back as 1929 by the Hilton–Young Commission but never implemented was intended as an advance over the previous concept of 'trusteeship' suggested by the Devonshire declaration, and it was assumed that henceforward self-government for East Africa, particularly Kenya and Tanganyika, would assume the form of a partnership between the different races which hitherto had lived in virtual social and political isolation from each other, despite their economic interdependence. However, because of the strong political and economic position of the Europeans in Kenya coupled with indecision and vacillation by the British government, the implementation of the policy of partnership there took a long, tortuous and difficult course.

The slow pace of change caused African suspicions to grow, while the post-war trends towards self-government in Lowland Africa made the Europeans increasingly anxious about their future fate in Kenya. After 1951, the year in which the Gold Coast (now Ghana) was granted internal self-government, African political agitation in Kenya, led by the veteran politician, Jomo Kenyatta, now the country's president, became greatly intensified, resulting in the Mau Mau uprising which began in 1952. This uprising was described at the time as a reversion to tribalism among the Kikuyu of Kenya, but in fact it was simply a means adopted by them after years of oppression and frustration for exploiting tribal sentiments to fight the difficult political battle against entrenched white interests in the country in the absence of any clear constitutional remedies. It was largely as a result of the uprising that Britain at last recognised the need to concede to the Africans a fair and equitable share in the government of the country and in its economy.

In 1953 a Royal Commission was appointed by the British government to examine the political and social problems of the whole of East Africa. The report of this Commission, published in 1955, led to a drastic review of land policies in Kenya, the admission of selected African farmers into the White Highlands and ultimately to the granting of self-government to the country on the basis of universal adult suffrage. At first it was sought to make political representation communal, but by the time independence was actually granted it had become clear that such a formula would be unworkable.

On the surface it seemed as if Tanganyika's progress towards independence would also be impeded by the presence of white settlers, who in 1960 numbered 22,300 as compared with 67,700 in Kenya and less than 10,000 in Uganda. Besides, after the Second World War deliberate attempts had been made by the British government to encourage the settlement of white immigrants in the country as a means of stimulating the economy, and the area allocated to non-Africans had been increased from about 809,360 ha (2 million acres) in 1939 to approximately 1,214,000 ha (3 million acres) in 1953.

However, because of the circumstances in which Tanganyika came under British rule and the fact that the European settlers were comparatively few in number and less compact in their distribution, the country was able to advance far more quickly and smoothly towards independence than Kenya, even though it was not until 1945 that the first Africans were nominated to the Legislative Council. But the country's immense size and the great variety and complexity of its environmental problems have made the process of nation building far more difficult than has been the case in either Kenya or Uganda.

Of the three East African countries Uganda is far the best endowed in terms of natural resources, which include copper, tin, coffee, cotton and tea. Also, it has always been a predominantly African country with only a small number of both Europeans and Asians. By any account, it had the best prospects for the easy achievement of independence. Unfortunately, as has happened in so many other African countries, in the final stages of the process leading to the transfer of power serious difficulties began to be created by petty differences and animosities between the various African kingdoms in the country harking back to past inter-tribal quarrels and conflicts. Right from the beginning of the country's contact with Europe, the largest and most developed as well as the richest kingdom had been Buganda, which stood far ahead of the three other kingdoms of Bunyoro, Toro and Ankole. Each of these kingdoms had a quasi-parliamentary system but lacked the ministerial system and cohesion of Buganda. The King of Buganda, the Kabaka, was consequently an important political figure, and independence for Uganda could not therefore come about until his agreement had been secured. In order to secure this agreement it was felt necessary by the British to grant the Kabaka a special position in the new government of the country. Thus, when Uganda eventually became independent, it had a popularly elected Prime Minister but a traditional and hereditary President, who was the Kabaka. Excellent as this unique arrangement seemed at the time, it contained the seeds for future discontent not only in respect of the relations between the Prime Minister and the President but also between the four rival kingdoms in the country, and until the final overthrow of the Kabaka in 1966 by a military coup organised by the Prime Minister much of the political life of the country revolved around the contest for supremacy between the occupants of these two principal offices in the country.

Tanganyika achieved independence on 9 December 1961 and assumed the new name, Tanzania, on becoming united on 26 April 1964 with the islands of Zanzibar and Pemba, which had been granted independence on 10 December 1963. Uganda followed on 9 October 1962 and Kenya on 12 December 1963. Many of the Europeans and Indians in Kenya had feared that the transfer of power to an African government would result in their wholesale ejection from the country, thus depriving them of the fruits of many years of hard toil. Thanks, however, to the wise statesmanship of the new African rulers, this did not happen – at least not as suddenly and as brutally as had been expected. The main change which took place in the economic scene was that increasing numbers of Africans gained an entry into the coveted highlands and gradually into other profitable sectors of the economy which had previously been the preserve of Europeans and Indians.

CLOSER UNION IN EAST AFRICA

Taken as a whole, East Africa is not a particularly rich region, and the process of providing it with adequate infrastructural services has never been an easy matter. It was therefore natural that after the First World War, when Britain assumed control of the whole region, the question of a possible union of the countries within the region should have been given serious consideration by Britain as a means of consolidating the administration and rationalising the various services needed for the effective development of the region instead of leaving each country to fend for itself on its own limited resources.

The first move for closer union took place in 1924 as a result of a motion in the British Parliament, which appointéd a Commission to report on the coordination of policy in both East and Central Africa, that is, the area covered by Kenya, Uganda, Tanganyika, Nyasaland, Northern Rhodesia and Zanzibar. The Commission, however, found little local support for the idea of a union and therefore recommended instead only the holding of regular inter-territorial conferences among the governors of the six territories and the heads of technical services for the purpose of discussing general policy matters and other subjects of common interest, such as agriculture and education (Hailey 1956).

In 1929 another Commission, known as the Hilton–Young Commission, appointed to examine the matter further, submitted its report. Again, the idea of union met with opposition from the white settlers in the region, and the British parliament decided therefore that the time was not yet opportune for such a move and proposed instead that there should be regular conferences of the governors of Kenya, Uganda and Tanganyika, with periodical extraordinary conferences involving in addition the governors of Northern Rhodesia and Nyasaland and the resident of Zanzibar, and that a joint Secretariat should be established for the purpose of considering technical subjects. The preference of the local European population was for the early establishment of a Union of Kenya and Tanganyika, but this was not accepted by Britain, which, as in the years immediately following the First World War, still felt doubtful about the propriety of involving Tanganyika, which was still a Mandated Territory, in a union with a colony fully under British control.

During the Second World War a number of bodies, including an East

African Production Supply Council, were set up by the British government to coordinate the economy and manpower of the East African territories, and as a result of this it was proposed in 1945 to attempt some practicable measure for dealing with services common to the three territories concerned but without actually attempting a political union such as had been proposed on previous occasions.

Apart from the problems created by Tanganyika's special status as a Mandated Territory, the fact that social and political conditions in each of the countries which were to form the proposed union, especially the power wielded by the local European communities and their attitudes towards the question of African rights, showed wide differences from one country to another made it extremely difficult to envisage a successful political union embracing all of them, and the British government was accordingly hesitant about taking any definite action.

It was therefore not until 1947 that the scheme for the coordination of common services put forward in 1945 received official approval; and even then it was confined to the three East African countries. Two bodies were created: an East African High Commission, which held its first meeting in Nairobi in 1948, and an East African Central Assembly with a provisional term of seven years expiring at the end of 1955. The High Commission was granted power to legislate, with the consent of the Assembly, in respect of a defined range of inter-territorial services, subject to the consent of the legislatures of the three territories concerned, on any matter concerned with the order and good government of the territories and was assigned specific responsibility for the administration of the railway, harbour, posts and telegraphs, and customs and excise services, as well as research and scientific services dealing with agriculture, forestry, health and fisheries. In addition to these matters it was also given responsibility for organisations dealing with two acute problems of an obviously regional nature, namely, tsetse and locust control.

The achievement of independence by the three East African territories in the early 1960s changed the whole basis for these arrangements; but the idea of intra-regional cooperation was not completely abandoned since all the new governments recognised the need for the maintenance of certain common services. In place of the East Africa High Commission and the Central Assembly a new body known as the East African Common Services Organisation was created in 1961 under the three Heads of State to handle the administration of the common social, scientific and economic services which the High Commission previously controlled.

Despite a number of initial difficulties and various conflicts which later arose within the organisation on account of ideological differences between the three member countries, especially Tanzania and Uganda, the organisation served the three countries well while it lasted. But it is now quite clear that the possibility of a political union of the three countries, as envisaged by their leaders immediately before and after independence, is now completely out of the question and even the organisation is in process of disintegration.

The staunchest advocate of such a union was President Nyerere of Tanzania, who, recognising the distractions that the separate achievement of independence by each country might produce, tried unsuccessfully to prevail upon Britain to speed up the independence of Kenya and Uganda and offered to delay Tanzania's independence in order to ensure that all three countries became independent simultaneously under a definite arrangement for the creation of a political union. For a number of reasons, including the lack of support from Britain and the fact that about the same time strong moves were afoot elsewhere in Africa for the establishment of a political union of the whole continent which had no place for regional groupings not specifically geared to the continental scheme, these efforts were not successful and in the event each of the three countries achieved its own independence separately.

Since independence the three countries have tended to drift increasingly apart, each of them insisting firmly on remaining master of its own house. However, there were certain spheres, such as transport and communications, in which cooperation was found to be not only desirable but absolutely essential in view of the imperatives of geography and location. Certainly, Uganda, which is a landlocked state, cannot possibly function as a viable political entity without being able to use the facilities of the Kenyan port of Mombasa or the roads and the railway through Kenya which give it access to the Indian Ocean. As between Kenya and Tanzania, also, an important human bond exists in the Masai people who straddle the political boundary that was arbitrarily drawn across the lands of the Masai by Britain and Germany at the end of the last century during the partition of East Africa. The idea of the East African Common Services Organisation was to give formal recognition to the common interests and needs of the three countries and provide the necessary machinery for meeting them. While such an arrangement was by no means the only way of meeting the needs of the countries concerned, there was certainly a great deal to commend it, especially in the early days of independence when none of the countries had the financial or human resources to meet its own infrastructural requirements and any attempt to share out what was already on the ground would have resulted in the dismantling or destruction of services and facilities that were already in effective operation. But the successful working of formal arrangements of this kind demands a considerable measure of goodwill and tolerance on the part of all the partners, and it is the gradual erosion of this essential foundation as a result of growing political differences between the three countries that has now led to the collapse of the Organisation.

But if the divisive forces of nationalism have tended to split up East Africa as a whole, within each country also sharp tribal differences have since independence created centrifugal tendencies that have made the process of nation building extremely difficult. Mention has already been made of the internal tribal differences that have plagued Uganda in recent years; in Kenya also political life since independence has been seriously undermined by differences and rivalries between the two principal tribes, the Kikuyu and the Luo. Tanzania has over 120 tribes, of which the largest is the Sukuma tribe, although fortunately it comprises only 13 per cent of the population and is therefore not in a position to dominate the political life of the country. But Tanzania has one great advantage over its two neighbours in that the *lingua franca* of East Africa, Swahili, is almost universally spoken in the country, thus providing an important basis for national unity.

Zanzibar had a serious racial problem before independence because of the concentration of political power in the hands of the Arab minority in the island. However, following a bloody revolution which took place after independence, power passed into the hands of the Bantu population; and the power of the Arabs has been further reduced by the political amalgamation of the island with Tanganyika in 1964, even though Zanzibar enjoys almost complete political autonomy with regard to its internal affairs. One of the factors no doubt responsible for holding the two parts of the country together since their union in 1964 has been the arrangement whereby the head of Zanzibar's government is automatically the First Vice-President of Tanzania, while the President and Second Vice-President are from the mainland. But the decision announced early in February 1977 to merge the mainland's only political party with that of Zanzibar is likely to strengthen the union further by giving it a common ideological framework based on the philosophy of *Ujamaa* or communal self-help and African socialism, which has been the main instrument of economic development on the mainland, especially in the rural areas, since the idea was formally launched by Tanzania's President, Julius Nyerere, in his famous Arusha Declaration of 1967.

In addition to these human problems practically every one of the East African countries is faced with a number of serious economic problems arising from the general inadequacy of their natural resource base and the resulting difficulty of mobilising the resources required for modernisation and social improvement, which are a necessary condition for the achievement of lasting political peace and stability within the region. But if the region's tangible economic resources in the form of minerals and agricultural products are limited by comparison with other parts of tropical Africa, it commands a highly valuable and unique asset in its natural scenery and wildlife which with proper management and exploitation could provide a steady and inexhaustible source of income for development in all the essential sectors. In the development of the resources available to it each country, most notably Tanzania, has adopted a different approach based on its own conception of economic and social development, and in the process the differences between them have tended to become even sharper than at the time of independence. Nevertheless, they still share many fundamental geographical characteristics, and the benefits that could accrue to all of them from closer cooperation with each other remain considerable.

STRATEGIC CONSIDERATIONS

In the days of the British Empire when Britain held important possessions not only in Africa but in and around the Indian Ocean, East Africa was a region of considerable strategic importance. The position has changed drastically since the decolonisation of the Empire and Britain's deliberate withdrawal from the area east of Suez. Nevertheless in the present air-age, East Africa has assumed a new importance as a stopping point on routes between Europe and Southern Africa and between America and Asia. Besides, its ports, especially Mombasa and Dar es Salaam, are still useful for tankers and other ships plying between the oil producing countries of the Middle East and European and American ports via the Cape of Good Hope or between Europe and various countries and islands within the Indian Ocean and along the western Pacific.

Lastly, the fact that there are still quite considerable numbers of British settlers in the region as well as substantial investments of British capital in one form or another gives Britain an intense but quite legitimate interest in the political and economic stability of the area and the maintenance of a climate generally favourable to the promotion of British commercial and financial interests.

I O

North-East Africa

North-East Africa consists of the Sudan and the three countries forming the 'Horn of Africa': Ethiopia, Somalia, and the Territory of the Afars and Issas, now known as the Republic of Djibouti (Fig. 28). The delimitation of the region is not altogether easy because it excludes Egypt, which, though undoubtedly part of North Africa, lies in the north-eastern portion of the continent and shares many close physical, cultural and historical affinities with the area, especially the Sudan, whose dependence on the Nile is exceeded only by that of Egypt, which controls the river's lower section. Nevertheless, Egypt's present-day political links with the rest of North Africa are so strong and vital that, even at the risk of being somewhat arbitrary, it is best on practical grounds not to include it within North-East Africa. Despite the fact that, as shown in the table below, the four countries of the region cover an area of nearly 4.4 million sq. km (1.7 million sq. miles), they contain a total population of only 49 million, of which as much as 28 million are to be found in Ethiopia, while the Sudan, whose area of nearly 2.5 million sq. km makes it the largest country in Africa, has a population of only about 18 million.

Political divisions	Area		Population
	Sq. km	Sq. miles	1975 or latest
Ethiopia	1,221,900	471,776	27,946,000
Djibouti			
(formerly Territory of the Afars and Issas)	21,783	8,410	200,000
Somalia	637,657	246,200	3,170,000
Sudan	2,505,813	967,494	17,757,000
Total	4,387,153	1,693,880	49,073,000

Geographically, the region falls into three clearly defined parts: (1) the middle and upper basins of the Nile on the west, with a distinctly low elevation and largely desert in character except for the southern half where the annual rainfall exceeds 254 mm (10 inches) and the vegetation changes steadily from desert scrub to savanna and finally to equatorial forest (in Uganda) along the Equator; (2) the high and rugged massif of Ethiopia in the centre consisting of formidable moun-

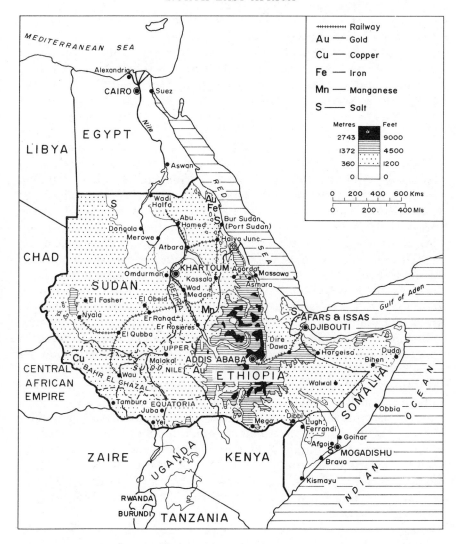

Fig. 28. North-East Africa: relief, political divisions, railways and minerals

tain ranges and plateau blocks carved out of thick beds of basalt overlying the ancient basal complex of Africa, with altitudes ranging from 1,000 m to over 3,300 m (3,300–10,000 ft) above sea level and traversed in the middle by the northern section of the East African Rift Valley; and (3) the arid and relatively low-lying plateau borderlands of the Red Sea and the Indian Ocean, where the annual rainfall drops sharply from between 1,270 and 1,778 mm (50–70 inches) in the central parts of the Ethiopian highlands to less than 508 mm (20 inches) with a corresponding deterioration of the vegetation from verdant mountain woodlands and pastures into scrub and semi-desert.

Because of its close proximity to the Arabian peninsula, from which it is separated only by the Red Sea and the narrow Bab-el-Mandeb Strait, and the fact that it lies at the meeting point of several different cultures representing North, West, Central and East Africa, the region contains a diversity of racial and ethnic types and forms a human and cultural bridge between Africa and the Middle East. Indeed, such is its physical and human diversity that it is difficult to think of it as forming a single geographical region, and it is more appropriately described as an ensemble of political divisions united principally by their common location within and around the Horn of Africa, their common frontage on the Red Sea and the Indian Ocean and the common exposure which all of them have had to Islam and Arabisation as a result of past invasions from the Arabian peninsula and the lower Nile valley.

With the exception of Ethiopia, where a wide variety of minerals such as gold, silver, potash, platinum and salt are known to occur in isolated pockets, the main economic wealth of the region lies in its agricultural resources. But even these are severely limited by the low and uncertain rainfall from which most of the region suffers. Only in the central highlands of Ethiopia where the rainfall is high and in the Gezira region of central Sudan where abundant supplies of water for irrigation are available from the Blue and White Nile does arable farming assume any real importance. In the greater part of the area conditions are so dry that the principal activity is pastoralism based on cattle, sheep, goats and, in the truly desert areas, camels. To a large extent the availability of water is the main determinant of population distribution, hence the high densities found in the Ethiopian highlands and in the central and southern parts of the Sudan. In the case of the Ethiopian highlands another factor making for high densities has been the protection against external invasion that has been offered through the years by the rugged and difficult nature of the terrain.

EVOLUTION OF THE PRESENT PATTERN OF STATES

The beginnings of the present pattern of states in North-East Africa and partly also of the present ethnic pattern can be said to date from the seventh century AD with the Arab invasion of Egypt and the Nile valley as far south as Nubia in present-day Sudan. The Arabs dislodged a number of powerful Christian kingdoms which, like the ancient Christian kingdom of Ethiopia, had been established for a number of centuries in this part of the continent but were prevented from penetrating into the southern part of the Sudan by the swamps and intricate channels of the Sudd which greatly impeded movement along the Nile.

During the first half of the nineteenth century the Egyptians also conquered the northern section of the Sudan and ruled it for about sixty years until they were driven back by the Mahdists. In the latter part of the century the British who were then in control of Egypt joined forces with the Egyptian army and regained the area taken over by the Mahdists ostensibly to restore Egyptian rule but in fact in order to safeguard their own territorial interests in the area and ward off the Italians, the Belgians and, above all, the French from the upper reaches of the Nile, whose uninterrupted flow was absolutely vital to the existence of Egypt.

In keeping with their declared aim of restoring the Sudan to Egypt, the British decided to share the government of the country with the Egyptians under a condominium agreement signed in 1899 and proceeded under a series of bilateral agreements with France, Belgium and Italy to define the country's boundaries along the west, south and east with the territories under the control of these powers. The boundaries on the west and south with French Equatorial Africa and the Belgian Congo were fixed along the Nile–Congo watershed, while that on the north with Egypt was generally fixed along latitude 22°N; and on the east Italy, which had occupied Kassala just west of the Eritrean border in 1894, was induced to return it in 1897 even before the Anglo-Egyptian agreement was signed. With the country's borders more or less fixed, the British began the difficult task of actually pacifying the southern part of the Sudan and integrating it with the rest of the country – a task which had still not been fully completed before the Sudan declared its independence and received formal recognition as a sovereign country from Britain and Egypt on 1 January 1956.

East of the Sudan the predominant power up to the seventh century AD was Ethiopia, which then embraced both the highland area associated with present-day Ethiopia and the lowlands extending eastwards to the Red Sea. As a result of earlier invasions by Semitic conquerors from South Arabia the original Cushite inhabitants of the area had been largely replaced by a hybrid race which now forms the dominant racial element in Ethiopia. In their mountain stronghold these Semitic immigrants established a strong Christian tradition claiming lineage with the biblical Queen of Sheba which has lasted without interruption until the present day and was the principal initial attraction to the Portuguese in the fifteenth century in their bid to circumnavigate Africa and establish a link with Prester John, the fabled Christian king of Ethiopia.

For several centuries, thanks to its location within what was in effect an impregnable mountain fortress, Ethiopia remained secure from external invasion with the result, as Gibbon observed, that it became a hermit kingdom. However, with the rise of Islam in the seventh century

AD, the eastern lowlands were invaded by Moslem conquerors from Arabia who occupied the coastal areas along the Red Sea and the Indian Ocean and overran considerable portions of eastern and southern Ethiopia, thus cutting off the mountainous core of the kingdom from its main contacts with the outside world. But although these conquerors succeeded in establishing the Moslem religion in much of the eastern and southern parts of Ethiopia they failed to produce any organised Moslem empire in the area, and by the end of the nineteenth century the Ethiopian Emperor, Menelik II, and his immediate predecessor, Johannes IV, had regained most of the country's earlier losses and consolidated them into a formidable state.

Right up until the end of the nineteenth century, however, and even later than that the Ethiopians continued to be harassed by some of their neighbours as well as by the European powers with interests in the area. Between 1875 and 1895 they successfully halted in turn Egyptian and Sudanese encroachments on the northern part of their territory but were then presented by an even more serious threat from the Italians, whose seizure of the Red Sea port of Massawa from Ethiopia in 1885 effectively opened the era of the European scramble for Africa in this part of the continent.

Italy was by no means the first European power in the area. As far back as 1839 the British had captured Aden at the south-western tip of the Arabian peninsula opposite the Somali coast and had developed it as a naval base and coaling station for ships sailing to India and the Far East. To strengthen Aden's position further and secure a source of meat and other vital supplies for Aden the British acquired a protectorate along the northern portion of the Somali coast in 1886, thus gaining an important vantage point on the Red Sea route through the Suez Canal which had been opened in 1869 and in relation to the 'Horn of Africa' and the basins of the middle and upper Nile. In 1862 the French had also established a foothold at Obock on the northern shore of the entrance to the narrow Gulf of Tadjoura linking the Red Sea to the Gulf of Aden as the result of a treaty of friendship and assistance with the Sultan in control of the area and in 1884 had acquired a protectorate over the area. To forestall any further French territorial gains not only within the 'Horn' but within the upper Nile basin the British encouraged Italy to acquire the extensive but mostly arid area to the south of British Somaliland, which eventually became the Italian colony of Somaliland in 1905, as well as the area around Massawa on the Eritrean coast to the north, leaving the French in control of the tiny protectorate they had established around the port of Obock. Later the French transferred the capital of the protectorate to Djibouti which became the main port and the terminus of the railway linking Addis Ababa to the Red Sea.

By the end of the nineteenth century the whole of North-East Africa, with the exception of Ethiopia, had been effectively partitioned among Britain, France and Italy. But even so Ethiopia was not wholly secure; the Italians tried to extend their Eritrean protectorate inland at the expense of Ethiopia and even went to the extent of claiming that Ethiopia itself was an Italian protectorate until they were defeated by the Ethiopian army at the Battle of Adowa in 1896. In spite of this the three European powers in the area refused to abandon their designs on the country, and but for the rivalry among themselves and the natural protection against attack provided by Ethiopia's mountainous terrain these powers might have occupied the country. In the event, although they all recognised Ethiopia's sovereignty after the Battle of Adowa, they nevertheless divided it up into spheres of influence under a tripartite agreement signed in 1906. In return for accommodating these powers the Ethiopians received various forms of assistance from them, the most significant of which was the construction of the Addis Ababa–Djibouti railway by the French between 1894 and 1917. In 1923, Ethiopia's new emperor, Haile Selassie, in a final bid to secure the political integrity of his country gained admission for it in the League of Nations. But even this provided only a temporary respite, for in 1935 the Italians who had never forgiven the Ethiopians for the humiliating defeat they inflicted on them at Adowa and were anxious moreover to extend their colonial possessions as a means of solving their acute population problems at home launched a totally unprovoked attack on Ethiopia and occupied it within a few months, forcing the emperor to seek refuge abroad until 1941, two years after the outbreak of the Second World War, when the British defeated the Italian forces and made it possible for Haile Selassie to return to his kingdom.

SUDAN, THE NILE AND ETHIOPIA

Of all the physical factors that have been responsible for shaping the geographical character of North-East Africa and directing the course of political events in the region the Nile is without any doubt the most important. The influence of the Nile is not confined to the main river; it extends to the whole basin, including most notably the western highlands of Ethiopia whose heavy summer rainfall is responsible for supplying the greater part of the river's annual flow.

From its most remote source near Lake Tanganyika to the Mediterranean, where it enters the sea after flowing through Egypt, the Nile measures 6,698 km (4,160 miles) and its total basin covers an area of 2,851,310 sq. km (1,000,000 sq. miles), containing over 45 million people. The river is made up of three principal tributaries of which the largest is the Blue Nile which draws practically all its water from Lake

Tana in the northern highlands of Ethiopia and contributes four-sevenths of the river's total flow. Next comes the White Nile, which is the longest branch and supplies two-sevenths of the main river's flow. The White Nile derives its headwaters from Lake Victoria and Lake Mobutu Sese Seko (formerly Lake Albert) and receives downstream in the Sudan another tributary, the Sobat, which derives its water mainly from south-western Ethiopia. Finally comes the Atbara, which drains the north-western part of Ethiopia and joins the main stream 322 km (200 miles) north of Khartoum and supplies one-seventh of the Nile's volume (Hurst 1957).

The flow of these three main tributaries is of crucial importance to the supply of water available to the Sudan and even more so to Egypt, which does not contain any significant affluent and hardly receives any water from natural precipitation. Also, while the equatorial sources of the Nile are important, the presence of the Sudd in the area results in a great deal of evaporation, and the most important sources of the river's supply are therefore the three tributaries reaching it from Ethiopia, which together are responsible for five-sevenths of the total annual flow.

Recognising the importance of the Ethiopian sources of the Nile, the British, right from the outset of their rule in the Sudan, concluded a treaty with Ethiopia in 1902 prohibiting it from 'starving or drowning Egypt' (Bowman 1928). Subsequently, in 1929, by the Nile Waters Agreement of that year, Britain put forward a scheme for the equitable distribution of the waters of the Nile between the Sudan and Egypt, and in 1959 a further Agreement was concluded in anticipation of various projected works on the river aimed at generating additional hydro-electric power for Egypt and increasing the supply of irrigation water to both countries (Suliman 1971).

Egypt is much more dependent on the Nile than the Sudan because practically every drop of water that is required for human consumption, for agriculture and for industrial purposes there is derived from the Nile, since apart from the narrow coastal strip along the Mediterranean the country is wholly desert. However, in the case of the Sudan, the southern two-thirds of the country receives varying amounts of rainfall ranging from under 203 mm (8 inches) per annum around Khartoum to over 1,422 mm (56 inches) per annum along the southern border of the country, and the main importance of the Nile lies in the irrigation water it supplies for the commercial cultivation of cotton and other arable crops in the Gezira area between the Blue Nile and the White Nile immediately south of their confluence at Khartoum. But the fact that the Nile flows through both countries and any use made of its waters in one either for irrigation or for the generation of hydro-electric power affects its availability for use in the other makes it absolutely necessary for both countries to take each other's needs into account in exploiting the river's resources whether in the form of water or of silt brought down by the annual floods. Similarly, both countries fully recognise the fact that Ethiopia, as the source of the Nile's most important tributaries, virtually holds the key to the amount of water that is able to reach them. The Nile has thus become a crucially important factor in the maintenance of peaceful and stable political relationships between Egypt and the Sudan and between these two countries and Ethiopia.

INDEPENDENCE AND NATIONHOOD

Politically, Ethiopia stands apart from the rest of North-East Africa by virtue of its long record as an independent state dating from antiquity and the fact that it is the only country to have survived the modern European partition and colonisation of the region more or less intact except for the brief period between 1935 and 1941 when it was conquered and occupied by Italy. The Sudan also succeeded generally in escaping political dismemberment, although it was conquered and ruled successively by Egypt and Britain during the nineteenth century and the first half of the present century until it gained its independence in 1956. The portion of the region that suffered the greatest dismemberment and disruption by the colonial powers was the eastern coastal strip along the Red Sea and the Indian Ocean which was partitioned among Britain, Italy and France to form British, French and Italian Somaliland and the Italian colony of Eritrea.

Before the opening of the Suez Canal in 1869 this part of Africa held very little interest for Europe because of its relative isolation and the generally inhospitable character of its physical environment. It was the opening of the Suez Canal that really drew attention to it, especially the portion lying along the Red Sea. Within the context of the Middle and Near East, however, the region was by no means a backwater; for centuries it had been in active commercial contact with the Orient as a source of myrrh and frankincense derived from Somalia and of ivory, gold and other precious minerals from the lands of the interior.

Somalia formed an extensive region occupied by various tribes exhibiting considerable racial and cultural homogeneity but lacking any coherent political organisation embracing the whole of their lands. At the time of the European entry into the region in the latter half of the nineteenth century the coastal area consisted of a number of separate Arab sultanates, while the interior formed a vague frontier zone contested by the Somali tribes and the Ethiopians who, under Menelik II, were pushing the frontiers of their empire eastwards towards the sea. Farther north, along the Eritrean portion of the Red Sea coast, both Ethiopia and Egypt were vying for possession at the time of the

European arrival in the area, although it seemed that the Ethiopian claim had a longer historical basis dating back to about AD 950, when the ancient kingdom of Axum which had been founded in the area originally by invaders from the Arabian peninsula came under Ethiopian rule.

It was with the consent of Ethiopia that Italy gained a footing in Eritrea in 1889 by the Treaty of Ucciali which led to the formal proclamation of the territory as an. Italian colony the following year. Subsequently, as has already been pointed out, the Italians tried to extend their territory at the expense of Ethiopia and even to control Ethiopia itself. But even though the Ethiopians put an end to their designs by defeating them at the Battle of Adowa in 1896, they made another attempt on Ethiopia in 1935, using Eritrea as their springboard, and took over the country.

After the defeat of the Italians in North-East Africa in 1941 Eritrea as well and British and Italian Somaliland were placed under British military administration, and for the first time Italian Somaliland, which had been badly neglected in the past, became exposed to liberal and progressive policies in the fields of education and other social services, which had the effect of stimulating the rebirth of Somali nationalism. The British had hoped for the immediate creation of a single Somali state embracing their own and the Italian portions of the territory as well as the Somali area of Ethiopia, but the idea was rejected by the United Nations, which in 1950 handed Italian Somaliland back to Italy as a Trust Territory with the stipulation that it should be prepared for self-rule by 1960 (Lewis 1971).

Fortunately the Italians responded positively to the challenge and between 1950 and 1960 seriously set about to prepare the territory for self-government by strengthening its institutions and involving the local populace in the government and administration. Spurred on by these developments the British also made redoubled efforts to prepare their portion of Somalia for self-government and actually handed over power to the Somalis four days before the Italian portion achieved its independence on 1 July 1960. But the leaders of the two territories had previously agreed to form a union, and accordingly on the very day that Italian Somaliland became independent it united with the British portion to form a unitary state known as the Somali Republic.

Having achieved their independence, the Somalis were faced with two problems: the integration of the two portions of their country which had been under separate colonial administrations and the recovery of Somali lands which the European partition of the area had placed inside the boundaries of Ethiopia and Kenya. Neither of these problems was easy, but the latter was far the more difficult because of its highly contentious nature and its far-reaching political implications.

SOMALI IRREDENTISM

Considering the different approaches to government and administration which the British and Italians followed in their respective portions of Somalia, the different cultural traditions which they left behind and the fact that English and Italian were employed as the official languages in the two portions of the country, the Somalis have made commendable progress since independence with the political and social integration of their new state. This is due in large measure to the strong feeling of unity which has been one of the outstanding characteristics of the Somalis throughout most of their recent history and was the main force behind the decision taken by their leaders at the time of independence to unite the British and Italian portions of their country into a single state. As a logical consequence of this a great deal of the country's energies since independence has been devoted to the development of a Somali irredentist movement whose aim is the recovery of Somali lands and Somali populations that were placed inside the boundaries of the country's two largest neighbours, Ethiopia and Kenya, as a result of the European partition of the region during the late nineteenth and early twentieth centuries.

There is no doubt that the delimitation of boundaries by the European powers in this as in so many other parts of the continent was done in a highly arbitrary fashion, as is clearly shown by the large number of straight-line boundaries found in the area. Indeed, up till now the southern sector of Somalia's border with Ethiopia to the point of intersection with the Kenya border has still not been properly delimited. Efforts by the United Nations during Italy's trusteeship of the area to settle the issue by arbitration met with no success, and other subsequent efforts have proved similarly futile. This has led to the development of intense guerrilla activity by the Somalis within the areas under dispute and a serious straining of relations with both Ethiopia and Kenya which shows no signs of abating. But the feelings of minority discontent are not only on one side; in 1965 several thousand Somalis crossed over into Ethiopia from the former British part of the territory in order to escape discrimination from the Somali government.

Unlike its British and Italian counterparts to the south, French Somaliland has continued to develop on its own as a separate entity since the end of the Second World War. Under the reforms made in the French Empire after the war the territory became an 'overseas territory' of France in 1946. Then, in 1956, under the newly introduced 'loi cadre' it was granted internal autonomy. In 1958 when de Gaulle made his historic offer of independence to France's overseas possessions French Somaliland opted to remain within the French Union, and subsequently in a referendum held in 1967 to decide the country's future the majority of the population consisting of Afars, as opposed to the more advanced and more highly organised Issas who have Somali affiliations, voted against independence and in favour of continued association with France. The only substantial change that occurred was the renaming of the country as the Territory of the Afars and Issas. Thus up to the end of 1976 it remained one of the few African countries still under foreign rule, and despite considerable pressure both from within and outside to secure complete independence for it France seemed unwilling to relinquish its hold of the territory.

In recent years both Ethiopia and Somalia have laid claim to the territory on the grounds of prior ownership and ethnic affinity, although in the case of Ethiopia another and even stronger reason is the fact that the territory's capital and main port, Djibouti, is the terminus of Ethiopia's main railway from Addis Ababa and handles more than 60 per cent of its overseas trade. Towards the end of 1975 there were strong indications that important changes were imminent in the territory's political status and its relations with France. Following consultations between the territory's leaders and the French government, a dramatic announcement was made at the beginning of January 1976 to the effect that France had at last decided to grant full independence to the territory in the very near future. As a result of this development Ethiopia agreed to renounce its claims to the territory and it was expected that Somalia would follow suit, thus making it possible for this small but strategically important territory to maintain its territorial integrity and continue to serve the legitimate interests of its neighbours and allies. France needs the territory as a base on the Red Sea, while Ethiopia needs the use of the port of Djibouti, which is vital for its external trade.

The fate of the Italian colony of Eritrea took a rather different turn after the Second World War from what many Eritreans had been led to expect. Although by the peace treaty of 1947 Italy was deprived of the colony, the Allied powers failed to reach agreement on its future until 1952, when by a decision of the United Nations it was federated with Ethiopia. However, on 14 November 1962 the Eritrean Assembly, acting no doubt under pressure from Ethiopia which had always regarded Eritrea as an integral part of its empire, voted for the outright union of the two countries under a single government.

On 27 June 1977 the Territory of the Afars and Issas was formally granted its independence by France under the name, the Republic of Djibouti. Although this ended the long drawn out uncertainty regarding the territory's political future it did not unfortunately dispose of many of the thorny issues regarding its relations with its two bigger and more powerful neighbours, Ethiopia and Somalia, both of which appeared still to have territorial ambitions in the new republic despite earlier hopes that these interests would be abandoned. Certainly, the present state of unrest among the countries in the Horn of Africa is very disturbing, and the Organisation of African Unity will have to do something about it quickly if early peace is to return to the region. Unfortunately, because of the strategic importance of the region a number of external forces are interested in securing special advantages there, and the ideological differences between these forces are quite clearly among the factors responsible for the political conflicts and polarisation that have erupted in the area.

PROBLEMS OF INTERNAL UNITY IN ETHIOPIA AND SUDAN

Although Ethiopia and the Sudan suffered relatively little territorial dismemberment from the European partition of North-East Africa, both of them have had to contend with serious problems of internal strife and disunity on account of their great ethnic and cultural diversity, which is itself a reflection of their varied historical past, and the nature of their physical geography.

Throughout its history the imposition of a single centralised authority over the whole of Ethiopia has never been an easy matter because of the rugged nature of the topography, which has tended to divide the country up into a large number of isolated blocks of territory bounded by steep scarps and separated from each other by deep river gorges (Fig. 29). Even under the Emperor Menelik II, who created the modern Ethiopian empire in the late nineteenth century, and his successor, Haile Selassie, the country consisted of several virtually autonomous provinces and smaller subdivisions ruled by powerful officials and local chiefs, whose loyalty and support were essential for the full exercise of the emperor's authority over the empire as a whole. There are also a large number of ethnic and tribal groups; but under the highly centralised system of government which prevailed during the long reign of Haile Selassie there was little scope for the assertion of divisive tendencies on the part of any of them even though some were clearly more favoured than others (Clapham 1969). The principal groups are the Amhara and the Tigreans who are mostly to be found in the attractive plateau region in the centre of the country; the Galla, who are akin to the Somalis and occupy the southern portion of the country; and, finally, the Negroid and other minority groups who are scattered in the low-lying borderlands. For centuries the Amhara, many of whom are Orthodox Christians in contrast to the Galla who are mainly Moslems, have been the most dominant group both politically and economically, but the indications are that under the country's new rulers the basis of power will be more equitably spread among the various groups. It is

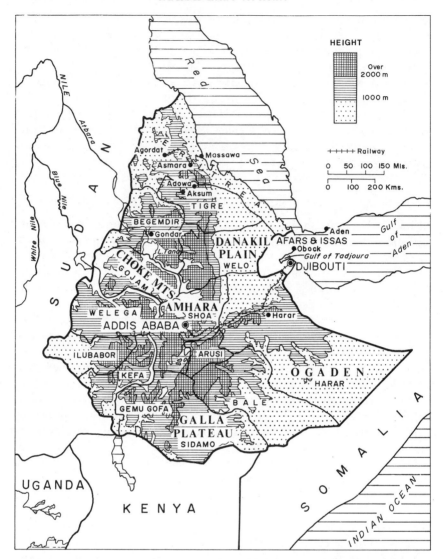

Fig. 29. Ethiopia: relief and internal regional and provincial divisions

noteworthy, however, that despite all the factors capable of generating discontent and internal division no serious moves for complete independence were made by any of the provinces until quite recently when Eritrea became an integral part of the country and a separatist movement began to develop there.

The Eritrean separatist movement has become an increasingly serious threat in recent years and can now be said to be at war with the new military government of Ethiopia which overthrew Haile Selassie in

1974. That Eritrea should want to be recognised as a separate state in its own right rather than remain just a province of Ethiopia is hardly surprising, considering that it was itself once the centre of a powerful empire which straddled much of the northern part of present-day Ethiopia and under Italian rule became separated from Ethiopia for some fifty years and acquired a new social, economic and cultural orientation. Besides, communication with the rest of the country is extremely difficult on account of certain formidable physical barriers which not even the widely acclaimed road-building skills of the Italians could effectively overcome during the period when Ethiopia was under Italian rule.

Apart from the Eritrean problem Ethiopia is faced with the long-standing problem of Somali minorities along its eastern and south-eastern border with Somalia, which, as has already been noted, has led to local guerrilla activity and a serious deterioration of relations with Somalia.

These two problems – the Eritrean separatist movement and the Somali minority problem – are now beginning to impose an unmistakable strain on Ethiopia's limited economic resources. It is important that they should be resolved as speedily as possible so that the country can attend to the many pressing social and economic problems that are crying out for solutions.

In the Sudan the cause of internal division is the marked racial and cultural differences between the peoples of the south and those of the north. While the northerners are predominantly Arab and adherents of the Islamic faith, the southerners are Negroes and Nilotic peoples with ethnic links across the borders of Zaire and Uganda, who during the period of British rule were kept virtually separate from their northern compatriots and exposed to the influence of European Christian missionaries. After the withdrawal of the British the new Sudanese government composed mainly of Arabs and Moslems tried to integrate them forcibly into the rest of the country by discouraging Christianity and insisting on the use of Arabic as the official language.

The effect of these sudden and generally unwelcome changes was the outbreak of a revolt by the southerners in August 1955, which was brutally suppressed by the central government, thus creating further discontent which finally led to a demand by the three southern provinces of Equatoria, Bahr el Ghazal and Upper Nile (Fig. 28, p. 219) for secession. The next important development occurred in 1963, when a violent underground movement known as the *Anya Nya* emerged, causing widespread terrorism, which in turn led to brutal military repression and the flight of several southern leaders into some of the neighbouring countries, especially Uganda (P. Kilber, 'Sudan', in Legum 1965). Although since then the Sudanese government has recognised the need for more moderate and conciliatory policies the hatred and suspicion created by these unfortunate events still remain and are unlikely to disappear for a long time to come unless some clear concessions are made to the southerners' demand for a federal form of government in which there will be greater scope for the expression of their ethnic and cultural identity.

These internal stresses are not really surprising when it is remembered that the Sudan's latitudinal range from about 5°N to 22°N is one of the greatest in Africa and that the country encompasses a wide diversity of physical environments ranging from wet equatorial and tropical conditions in the south, where the dissident groups are to be found, to extreme desert conditions in the centre and north, where the dominant Arab peoples are mainly concentrated. But the southerners are severely handicapped by the fact that they occupy only a small portion of the country amounting to not more than one-third, which is economically backward, while the central and northern parts where the Arab elements live are not only very much larger but contain most of the country's wealth.

As is to be expected, the southern problem led to strained relations between the Sudan and its southern neighbours, especially Uganda, who regard the southern peoples as their kinsmen and support their resistance to the process of Arabisation and Islamisation to which the Arab-dominated government of the Sudan seems determined to subject them. However, things have since improved considerably.

ECONOMIC AND STRATEGIC ASPECTS

What has been said so far is enough to support the assertion made at the beginning of the chapter that North-East Africa does not truly constitute a geographical region except in respect of its location in the north-east corner of the continent, within and around the Horn of Africa. More than that it is an area whose constituent political entities do not appear to have any common and unifying political and economic goals but, on the contrary, are torn apart by internal dissension and conflict which have become greatly accentuated with the attainment of independence.

Addis Ababa, the capital of Ethiopia, is the headquarters of the Organisation of African Unity; but the essential purpose of the Organisation as an instrument for the promotion of unity and harmony among the countries of Africa appears not to have made any significant impression on the region where it is based. Today, the region, more than any other region in Africa, epitomises the dangerous conflicts and stresses bequeathed by the European partition of the continent and the imposition of arbitrarily drawn boundaries upon its peoples.

What attracted the European powers to North-East Africa in the first instance was its strategic rather than its economic importance. These considerations still hold true today; and the two principal axes of the region, the Nile valley and the Red Sea and Indian Ocean coasts, especially the latter, are still centres of intense international interest – the Nile valley because it holds the key to Egypt's existence, and the Red Sea and Indian Ocean coasts because of their strategic location on the sea route leading to and from the oil-rich countries of the Middle East, whose attitudes to the rest of the world are now determined to a large extent by the political attitudes and moods of Egypt. Thus, in a sense, North-East Africa has become inevitably drawn into the complex and volatile politics of the Middle East by reason of its location and its cultural and historical affiliations.

Economically, however, it is a relatively poor region endowed with few except agricultural resources and a very limited base for industrial development. It is a well-known fact that most of the present systems of transport and communications in Africa were developed during the colonial period and that the areas which received the greatest attention were those with proven or clearly obvious economic potential. The marked absence of well-developed railways and roads in North-East Africa is a clear measure of the low importance accorded to it in economic terms by the colonial powers which exercised political control over the area. Road development is extremely poor in the region as a whole, particularly in the Sudan, despite its vast size, and only railways are available to handle the bulk of its west–east trade between the Darfur and Kordofan regions and the Red Sea ports of Bur Sudan (Port Sudan) and Sawakin. The Nile between Juba and Khartoum plays a key role in the country's internal south–north trade, while the southern regions of the Sudan are virtually isolated and cut off from the rest of the national economy. In Ethiopia the old railway from Addis Ababa to Djibouti remains the country's main lifeline, with Massawa and Assab on the Red Sea coast of Eritrea playing only a very minor role as compared with Djibouti, which before the closure of the Suez Canal in 1967, was handling 1,000 ships annually. Somalia is totally without railways and its principal port of Mogadishu has only road connections with the interior (Chi-Bonnardel 1973). Such a meagre transportation system obviously cannot serve the needs of all the four countries within the region, and more adequate systems will have to be developed by almost every one of them, especially Somalia, as part of the vital process of providing the essential infrastructural base for effective national development.

I I

Islands of the Indian Ocean

Located in the south-western portion of the Indian Ocean, between the Equator and the Tropic of Capricorn, are a number of islands and archipelagos that are regarded as an extension of the African continent. Politically, they fall into five divisions, as shown on the Table below: Madagascar, sometimes known simply as 'The Great Island' (Stratton 1965) but renamed since independence as the Malagasy Republic, the Comoros, Réunion, Mauritius and the Seychelles. They form a special world of their own, half-African, half-Asian, but linked to Africa by many cultural and historical ties (Fig. 30) (Chi-Bonnardel 1973). In physical and structural terms, however, only the Comoros and Madagascar are true extensions of the continent, the rest consisting of oceanic islands representing extinct or partially extinct volcanoes or standing on isolated continental fragments rising from the ocean floor.

Political divisions	Area		Population 1975 or latest
	Sq. km	*Sq. miles*	
Comoros	2,236	863	306,000
Madagascar (Malagasy Republic)	587,041	226,657	8,020,000
Mauritius	2,045	790	899,000
Réunion	2,510	969	467,675
Seychelles	376	145	60,000
Total	594,208	229,424	9,752,675

With the exception of Madagascar, none of these islands was inhabited until comparatively recent times, but, thanks to their location and their generally congenial oceanic climates, they have attracted a wide variety of immigrants from within and outside the Indian Ocean and their populations today consist of highly complex mixtures of diverse physical types and racial stocks drawn from Africa, Arabia, India, Malaysia, Indonesia, China and Europe, especially France and Britain which ruled them until quite recently when they achieved their independence. Madagascar was probably colonised round about the year AD 1000, but all the other islands received their first immigrants after their discovery by the Portuguese in the sixteenth century. Apart from

234

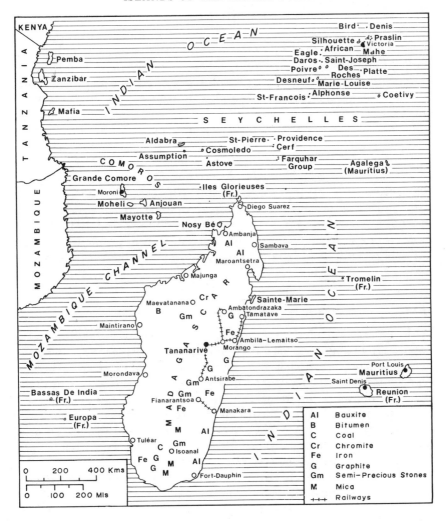

Fig. 30. Islands of the Indian Ocean, showing major railways and minerals in Madagascar

the Africans of Bantu stock who were mostly brought in as slaves from the eastern coast of Africa and the Cape Colony by the Portuguese and the Dutch, the other racial elements came in mainly as voluntary immigrants from the lands within and around the Indian Ocean, while the Europeans represent mostly the descendants of sailors and traders using the islands as convenient stopping places on the sea route to India round the Cape of Good Hope together with a few others, mostly in the Comoro islands, descended from deportees introduced by the French during the French Revolution.

Before the opening of the Suez Canal some of the islands, such as Mauritius, served as important coaling and victualling stations for ships plying across the Indian Ocean, but with the opening of the canal these functions were taken over by other places like Aden and Djibouti located more conveniently at the southern end of the Red Sea. However, all the islands continued to play an important role as strategic outposts for Britain and France and as sources of certain valuable tropical products like sugar, vanilla, cinnamon and other rare spices.

In the vastness of the Indian Ocean these islands appear to be located quite close to the African continent; but in fact the easternmost of them, the Seychelles and Mauritius, lie about 1,932 km (1,200 miles) from the nearest point on the East African coast. Also, although they share a number of common geographical characteristics, their individual social and economic conditions differ so widely from each other that in discussing the characteristics and problems of their political geography it is best to treat each one of them as a separate unit.

MADAGASCAR (MALAGASY REPUBLIC)

With an area of nearly 591,000 sq. km (228,000 sq. miles) and a population of slightly over 8 million, Madagascar is far and away the most important of the African islands of the Indian Ocean. Its area represents almost 99 per cent of the area of all the five islands within the group, while its population represents just over 83 per cent of their entire population. Among African countries it ranks as one of the largest in terms of area, while in world terms it has the distinction of being the fourth largest island on the globe after Greenland, New Guinea and Borneo (Fig. 30).

Once physically part of Africa, the island is believed to have become detached at the end of the Palaeozoic era, thus giving its flora and fauna a strikingly different character from those of the mainland, where the much more extensive geographical base available for subsequent evolution appears to have resulted in a much greater variety and complexity of both plant and animal types. Consequently, most of the rich and abundant life forms which constitute such an important natural resource in many of the tropical and sub-tropical countries of the continent are totally lacking in Madagascar.

Another consequence of the island's early physical separation from the African continent is that for several centuries it remained completely uninhabited by man; and it was not until during the Iron Age that the first inhabitants appear to have arrived from neighbouring parts of Southern Africa, followed by Indonesian and Polynesian navigators. Round about the ninth century AD Arab Moslems of diverse origins also moved in and gradually interbred with sections of the original population.

Unlike the majority of African countries, Madagascar suffered little interference from Europeans right up until the final decades of the nineteenth century. As far back as the sixteenth century the Portuguese and, subsequently, the Dutch, French and British had established coastal bases on the island, but this was more because of its strategic location on the sea route to India than because of any special opportunities it offered for trade. Apart from the absence of any very strong economic incentive for European colonisation and settlement there was the equally important fact that by the sixteenth century the inhabitants of the island had become organised into a number of fairly powerful tribal kingdoms which though very much divided against themselves were all united in their opposition to European colonisation.

The most powerful of these kingdoms was the Merina kingdom based in the central highlands of the island where the present capital, Tananarive, is located. It was composed of light-skinned people of Indonesian and Polynesian origin. By the nineteenth century the Merina kingdom, with the help of a few favoured British and French advisers who were allowed to settle on the island, had succeeded in extending its political hold over practically the whole of the island. Although the British were content to accept the political dominance and rule of the Merina in the island, the French had different aims and from 1883 onwards embarked on the outright conquest of the island. In 1890 the British, in return for concessions in Zanzibar, agreed to recognise a French protectorate over the island, and the French, now in full control, proceeded to abolish the Merina royalty which had by then degenerated into a cruel autocracy violently opposed to Europeans and the local converts of the European Christian missionaries.

The island was declared a French colony and by 1898 French rule had become firmly established. However, Malagasy nationalism continued to simmer beneath the surface, and even though in 1946 all the Malagasy were granted full French citizenship a massive rebellion broke out suddenly in 1947, which clearly pointed to the need for further political concessions aimed at complete autonomy. As in the case of the other French African territories apart from Guinea, the Malagasy responded to General de Gaulle's historic offer in 1958 by initially accepting internal autonomy within the French Union, but in 1960, like their mainland counterparts, they demanded full political sovereignty, and Madagascar thus became independent on 26 June 1960 under the name Malagasy Republic.

Since independence the Malagasy Republic has become a very active and vocal member of the Organisation of African Unity and has established close political and diplomatic links with the more moderate

African states formerly under French rule. The republic has also maintained close economic and cultural links with France despite the bitterness created by the struggle for independence. However, the course of events since 1960 has not been particularly smooth, partly because of the island's ethnic heterogeneity, but more especially because of economic difficulties.

There is no doubt that the island has considerable economic potential, including deposits of graphite, chromite, mica, coal, bitumen, iron and bauxite, but it is handicapped by the lack of an adequately developed infrastructure due to administrative neglect under French rule as well as by the absence of capital necessary for really effective development. It also suffers from the physical isolation imposed by its geographical location and the fact that its closest cultural neighbours in Africa are the former French colonies, all of which are located far away on the other side of the continent. In some respects also the small size of its population works in its disfavour by making it difficult to obtain enough local labour for economic development.

COMOROS, RÉUNION, MAURITIUS, AND THE SEYCHELLES

These four islands share many of the geographical, social and economic characteristics and problems that have been noted in the case of the Malagasy Republic, except that in their case the problems are made even more acute by their very tiny areas and the very limited material base available to them for development. Until quite recently the future seemed particularly grim for Mauritius, which had both the highest population density among the entire group (440 persons per square kilometre) and the highest population growth rate (3.12 per cent in 1962) (Benedict 1965). However, thanks to successful population control programmes during the past decade aided to a certain extent by emigration, the rate has now been reduced to 1.5 per cent per annum. The position of Réunion, the Seychelles and the Comoros with population densities of respectively 186, 160 and 137 persons per square kilometre is also quite serious although not as alarming as that of Mauritius before the introduction of its progressive population policies, which clearly deserve to be emulated by all the other islands (Fig. 34, p. 255).

The colonial affiliations of the Comoros and Réunion have been with France, while those of Mauritius and the Seychelles have been with Britain. All of them started primarily as strategic bases and neither France nor Britain gave serious thought to the possibility of eventual political independence for any of them even after the African independence movement got seriously under way because of their very tiny sizes and their precarious economies. Besides, owing to their strategic value neither the British nor the French were particularly anxious to speed their progress towards independence. All that seemed to be required in the view of the two powers was a reasonable liberalisation of the internal political institutions of the islands so that their inhabitants might have a greater say in the management of their own affairs.

Réunion

In the case of the French island of Réunion, the largest island of the Mascarene Archipelago, which became a French possession in 1649 and acquired the status of an overseas department of France in 1946, the policy followed by the French of confining political reform purely to matters affecting the island's internal administration while retaining all real power in their own hands was given decisive endorsement by the inhabitants when they voted in 1958 to remain within the French Union as a department of France. Since then this verdict has been endorsed by the majority of the population despite strong and persistent pressures on them from outside, notably from the Organisation of African Unity, to demand full political sovereignty.

Comoros

In the case of the Comoros, which consist of four small islands located between the northern tip of Madagascar and the coast of Mozambique on the African mainland, the trend of popular opinion on the question of independence has been quite the opposite. Apart from the pro-French island of Mayotte, which was the first to be colonised by the French in 1843, all the islands favour independence from France.

Local pressures for independence began seriously in 1965 with the backing of several thousand Comoro citizens living in Tanzania and in Madagascar, which was responsible for the administration of the Comoros from 1912 until 1948 when they achieved the status of a French overseas territory. In 1956 the French began a programme of gradual but somewhat half-hearted administrative reforms aimed at giving the islands a greater say in their internal affairs, but these failed to satisfy the nationalists who were not prepared to settle for anything short of complete political autonomy, and even though the political party for the liberation of the Comoros formed on the islands in 1964 was banned by the French a counterpart movement based in Tanzania continued to work and agitate for the same goal.

The most significant result of these efforts occurred in 1974, when the French at last agreed to hold a referendum on independence at the end of the year. Although the outcome of the referendum was decisively in favour of independence, the French, in deference to the wishes of the inhabitants of Mayotte, decided to offer the Comoros a constitution which, while granting independence to the other islands within the group would give Mayotte a special status – possibly that of a department of France just like the island of Réunion. The presumption is that this would enable the French to establish a new naval base there in place of their former base at Diego-Suarez at the northern tip of Madagascar, which the Malagasy government recently forced them to relinquish.

To forestall such a development the members of the Chamber of Deputies from the other islands, who far outnumbered those from Mayotte, made an unexpected unilateral declaration of independence on 6 July 1975 in the capital, Moroni, on the island of Grande Comore, thus precipitating a serious constitutional crisis which led to considerable confusion in the islands and the eventual deposition of the President in a coup while he was visiting the neighbouring island of Anjouan. Since the President had been responsible for the original declaration of independence, it was widely believed that France had been responsible for engineering his overthrow.

Although France has rejected the charge and has announced its preparedness to recognise the independence of the Comoros, the question of the future of the island of Mayotte still poses a difficult problem for which no clear solution appears to be in sight. It is quite obvious that unless this problem is resolved amicably there will be little chance of peace in the Comoros even after the formal granting of independence by France.

The root of the problem is that Mayotte is predominantly Christian, in contrast with the other three islands of Grande Comore, Anjouan and Moheli, which are largely Moslem, and the inhabitants of Mayotte who number only 37,000 out of the islands' total population of 289,000 are reluctant to place their political future in the hands of a government which is certain to be dominated by Moslems. At the same time, however, unless the Comoros are to seek new and equally helpful political allies after independence, it is in the interest of all the islands to maintain reasonably cordial relations with France, which has so far been responsible for propping up the territory's shaky economy by supplying about 80 per cent of its annual budget in the form of grants.

Mauritius, and the Seychelles

The two former British dependencies of Mauritius and the Seychelles form the easternmost of the African islands of the Indian Ocean and share many close historical associations with their French counterparts, especially Réunion. Like Réunion, both are volcanic in origin and were once ruled by France. But the geographical affinity between Mauritius and Réunion is especially close, for both belong to the Mascarene Archipelago and are separated from each other by a distance of only about 242 km (150 miles).

Mauritius

Like so many of the islands of the Indian Ocean, Mauritius was first brought to the attention of the world by the Portuguese in the sixteenth century, although it is probable that long before then the island may have been visited by Arab navigators. The Portuguese made no attempt at settlement and merely used it as a victualling station for their ships plying the Indian Ocean. In 1598 the Dutch took over the island and established plantations of sugarcane, tobacco and cotton with the help of slave labour imported mostly from their colony at the Cape of Good Hope. But they abandoned it in 1710 on account of dissensions among the settlers, leaving behind a few runaway slaves in the interior. The Dutch were followed by the French, who claimed the island in 1715 and used it as a base for operations against the British in the Indian Ocean and in India itself until its capture by the British in 1810 during the Napoleonic Wars. The Treaty of Paris in 1814 confirmed British possession of the island and its dependencies, Réunion and Rodrigues, a small island to the east within the Mascarene Archipelago, although Réunion was later returned to France.

During their occupation of the island the French also imported slave labour from Madagascar for the cultivation of the sugar plantations, which they greatly extended, and by the time of the British takeover had left a strong social and economic imprint on the island which still persists today in the large number of people of French descent and of Creoles representing people of mixed French and African, Malagasy or Indian descent, and in the large sugar plantations which now dominate the island's cultural landscape. Other crops, such as cotton, coffee, indigo and cloves, were also grown on the island, but sugar was given special preference because it was much less susceptible to damage by the violent cyclones which frequently sweep the island. After the abolition of the slave trade, which took effect on the island in 1835, the French sugar planters, most of whom stayed on, found it difficult to obtain local labour for their plantations and so resorted to the importation of indentured labourers from India, just as the British were doing in Natal about the same time. It is from these labourers that the large number of Mauritian Indians, forming over 50 per cent of the island's present population, are descended. A few Chinese were also introduced into the island by the French, but they remained mostly a small and closed community rarely intermarrying with other races, especially the Europeans, and generally maintaining close links with their original homeland. With the establishment of British rule came a number of British, but very few actually settled permanently, the vast majority remaining only as temporarily resident administrators.

It was during the period of British rule that Mauritius fully acquired its present character as a plural society made up of Franco-Mauritians, Creoles, black inhabitants derived from Malagasy and African slaves, Indians comprising Moslems and Hindus, Chinese and British. Each of these groups occupied a more or less distinct position in the island's economic and social hierarchy, with the Franco-Mauritians and the British at the top and the Indians and descendants of the original black slaves at the bottom.

Under British rule little was done to change the island's essential social and economic structure based on the exploitation of the labour of the large majority of the population by a handful of rich, land-owning families and individuals, mostly of French origin, who controlled the lucrative sugar industry. Political power remained almost exclusively in the hands of the Franco-Mauritians and the Creoles, while the large Indian population had little say in the conduct of affairs. On the surface the island seemed peaceful, but beneath there was widespread discontent.

This discontent finally erupted in 1937, when widespread riots broke out among the labourers and small planters on the sugar estates, thus forcing the British to introduce a number of long-overdue political reforms based on the recognition of the rights of the large non-European section of the population. For the first time elections were conducted on the basis of a wide franchise which clearly favoured the Indians and the Creoles and greatly reduced the representation of the Franco-Mauritians in the island's legislature.

These reforms paved the way first for internal self-government and ultimately for independence. Throughout these developments, however, there remained the difficult question of the island's future relations with Britain, on which it depended for most of its trade and financial support, coupled with the natural fear of the European population regarding their future in an independent Mauritius dominated by non-Europeans.

It was under the cloud of these uncertainties that Mauritius was granted its independence by Britain on 12 March 1968. But there was an understanding with the new government that Britain would continue to maintain air and naval bases on the island and, even more important, offer automatic military assistance in the event of a violent overthrow of any popularly elected government.

Despite these assurances, the future of the island is by no means secure, largely because of the very limited and fragile basis of its economy and the inherent instabilities in its social structure, which tend to militate against that widespread sense of unity essential for the development of true nationhood. The economy is almost exclusively based on sugar, which covers 85 per cent of the cultivated land and accounts for 90 per cent of all exports or 99 per cent if sugar by-products are included. It is estimated that sugar and its by-products alone account for more than one-third of the total national income. Yet because of the island's proneness to violent tropical cyclones the possibility of developing other crops on a large scale is almost completely ruled out, although coffee and tea are still grown in specially favoured locations.

Another serious problem is the inequitable distribution of land. As a result of the long history of plantation agriculture controlled by a few wealthy families and individuals, as much as 55 per cent of the land under sugar is owned by a small handful of Franco-Mauritians, while the remainder is cultivated by some 84,000 freehold planters, whose holdings range between approximately 0.3 and 203 ha (1–500 acres) (Benedict 1965). The vast majority of the population are completely landless and are consequently obliged to seek wage employment, which is not very plentiful. The problem is further compounded by the high population density, of over 400 persons per sq. km, although the recent reduction of the rate of population growth from 3.12 per cent per annum in the early and middle 1960s to 1.5 per cent per annum should help considerably in improving the position. It is thus faced with a very bleak economic future, which can only be ameliorated by resort to a policy of stringent population control coupled with a really determined effort at economic diversification.

The expansion of tourism offers a partial solution; but this depends among other things on the extent to which a reasonably stable social and political climate can be maintained. In the initial stages of independence there seemed to be a possibility that aid might also be forthcoming from Britain, especially in view of the air and naval bases which it was to maintain on the island, by mutual agreement, for an initial period of six years. But this prospect has now diminished considerably, following Britain's decision taken in 1975 to close these bases and end the defence agreement concluded with the island at independence as from the end of March 1976 in response to pressures from the Mauritian government. This development represents an important shift of policy on the part of the Mauritian government; for there are strong indications that the vacuum created by Britain's withdrawal will be filled by the Soviet Union, which over the past few years has been strengthening its naval presence in the Indian Ocean and has already acquired a number of bases and other important military facilities in the area.

Surprising as this shift in policy may appear, it goes to underline the dilemma in which Mauritius now finds itself in the conduct of its external affairs. In the rapidly changing balance of power situation in the Indian Ocean it is quite apparent that Mauritius is not an entirely free agent in the determination of its political alignment. As a small island saddled with serious internal problems and cut off from the rest of the world by vast distances, it stands in desperate need of allies who can be relied upon both militarily and politically for protection and support in time of need. From both points of view its best hope lies in the establishment of the closest possible links with the African countries which are its nearest geographical neighbours as well as with those countries or power blocs which clearly exercise a dominant role in the Indian Ocean. Since independence the island has been a member of the Organisation of African Unity and in 1970 it also joined the Afro-Malagasy Common Organisation (OCAM), which has since been renamed the Afro-Malagasy and Mauritian Common Organisation, (OCAMM). In order to maintain these links and enjoy the support they are able to offer it is incumbent on Mauritius to avoid any foreign alignments that do not generally command the approval of its African neighbours, a number of whom have tended to favour political alignments with the Eastern rather than the Western power bloc.

But even more serious than the question of the island's external relations is that of its internal stability. Throughout the period of colonial rule no serious effort was made to integrate the diverse ethnic groups, who were sharply divided from one another socially and politically by narrow communal interests, even though they were all economically interdependent in the manner of a typical plural society.

Now that independence has been achieved and political power has become almost wholly dependent on numerical size and popular support, which clearly favour the large Hindu section of the Indian population, the most serious problem facing the island, next to its economic difficulties, is how to knit its diverse ethnic elements together into a single coherent Mauritian nation united by a common loyalty transcending the narrow interests or claims of any particular group. Considering the numerical predominance of the Hindu population, only tolerance, moderation and wise statesmanship on the part of the government can provide a lasting solution to the problem.

Seychelles

The Seychelles consist of an archipelago of some ninety small islands and cays of coral and granitic origin standing on what is believed to be a detached portion of the ancient continent of Gondwanaland and strewn over a distance of about 1,207 km (750 miles) from east to west. The largest island of the group, Mahe, on which the capital, Victoria, is situated, has an area of 143 sq. km (55 sq. miles) and lies about 1,610 km (1,000 miles) east of Mombasa on the Kenya coast, 1,497 km (930 miles) north of Mauritius, a little over 966 km (600 miles) north of Madagascar, and 2,818 km (1,750 miles) south-west of Bombay on the Indian sub-continent. Located within only a few degrees of the Equator and devoid of any very high altitudes, the archipelago has a typically hot and humid tropical climate, which is considerably moderated by oceanic influences.

As in the case of Mauritius, the Portuguese were the first Europeans to visit the islands early in the sixteenth century, although they are believed to have been visited as far back as the twelfth century by Arab and Persian navigators. In 1609 they were also visited by a British ship; but it was the French who first settled on them in 1768 and established plantations for spices with the assistance of slaves brought over from Africa. However, in 1794 the British took over and for the next twenty years the islands changed hands several times between the two nations until they were formally ceded to Britain in 1814 by the Treaty of Paris.

Like the French before them, the British administered them from Mauritius, which was by then in British hands, but the islands were given a separate administration as a Crown Colony under their own governor in 1903. By then the spice plantations, which the French had deliberately destroyed in the eighteenth century to prevent them from falling into the hands of the British, had virtually become extinct and the economy of the islands centred on the production of copra, with spices playing only a very secondary role. But the economic value of the islands to Britain remained negligible and their main importance lay in their occasional use for the detention of troublesome opponents of British rule in various parts of the British Empire, such as King Prempeh I of Ashanti, who was sent there with leading members of his Court, including the intrepid woman warrior, Yaa Asantewa, in the early part of the present century and, most recently of all, Archbishop Makarios of Cyprus, who was detained there from 1956 to 1957.

The Seychelles is a very poor country with very limited economic resources apart from copra and a few spices, especially cinnamon. Of the ninety or so islands that comprise the archipelago only two, Mahe and Praslin, are permanently inhabited. Most of the other islands are covered with coconut palms which have sprung up through natural propagation, but they are so distant from the two permanently inhabited islands that their regular exploitation is impracticable.

Because of its very limited economic base the Seychelles has attracted few immigrants from outside and the present population consists mainly of the descendants of the original French settlers and African slaves together with a sprinkling of Asians from China, India and Malaya who arrived later as traders, all of whom have interbred over the centuries to produce a highly mixed race of Creoles with many gradations of skin colour speaking a language which is a creole patois made up of archaic French words intermixed with words from various African dialects and a few mutilated English words. The official language is English, but French is also spoken quite widely.

Until quite recently the main concern of the British administration and of the people of the Seychelles themselves was the economic improvement of the islands, and the question of political independence was hardly ever mentioned. It first became an issue in 1965 when some of the islands' political leaders put it forward as their goal. Surprisingly, however, the idea was strongly opposed by the vast majority of the population, whose preference was for an association with Britain similar to that enjoyed by the Channel Islands. Britain, on the other hand, favoured independence and thus found itself in the highly embarrassing position of wanting to impose self-rule upon an unwilling colony. But in an election held in 1970 under a revised constitution aimed at giving the colony internal self-government in preparation for eventual independence the people again reaffirmed their previous stand and elected to power the party opposing independence.

As was to be expected, these developments did not find favour with the Organisation of African Unity, which openly supported the opposition party pledged to the achievement of independence, and the Organisation exerted considerable diplomatic pressure on Britain to end its rule of the colony. Faced with what was clearly an untenable situation, the party in power reversed its earlier stand and decided in 1974 to ask Britain to grant independence to the Seychelles. Following constitutional talks in London in March 1975, the colony was granted independence on 28 June 1976.

True to the pattern in the other African islands of the Indian Ocean, the Seychelles government has since joined the Organisation of African Unity and declared its opposition to the establishment of foreign bases on the islands. Meanwhile Britain is pressing on with a number of essential development projects aimed at improving the internal communications of the islands, equipping them with a modern airport capable of serving the biggest jets plying on international routes across the Indian Ocean, and at providing an adequate infrastructure for an expanded tourist industry which can bring in much-needed foreign exchange. But the country will continue to need outside financial assistance, which Britain has hitherto provided, in order to balance its budget if it is to survive as an independent state; and it was no doubt fear of the economic uncertainties that might follow independence that led to the long opposition of so many of the inhabitants to the idea of independence from Britain.

In contrast with the gloomy economic outlook, the social climate of the Seychelles is much more encouraging and augurs well for the future. Although there is unmistakable social stratification with white landlords at the apex of the social pyramid and landless labourers descended from slaves at the base, nevertheless as a result of the many centuries of miscegenation the population is generally well integrated and the divisive tendencies usually brought about by the presence of self-conscious minority groups or strong sectional interests are not particularly important (Grove 1967). Thus the country can claim to have at least one of the main prerequisites of nationhood – a united population conscious of a common identity.

As far as external political affiliations are concerned, it is very likely that the Seychelles will follow the example of Mauritius and seek its political support mainly from the Organisation of African Unity and those foreign nations and power blocs that are acceptable to the majority of its members. Whatever the intentions of other foreign countries might be, it is unlikely that either Britain or the United States will press for bases on the islands since they have already taken other steps to meet their needs. As far back as 1970 the two countries, possibly in anticipation of likely shifts in policy by the newly independent island states in the western part of the Indian Ocean on whom they had previously relied for strategic bases, decided to construct a naval communications base on the island of Diego Garcia in the Chagos archipelago, formerly administered by Mauritius, in order to fill a serious gap in the United States' global communications systems, provide support for American and British ships and planes in the Indian Ocean, and also serve as a facility for monitoring the movements of the expanding naval presence of the Soviet Union in the Indian Ocean. In preparation for this Britain had decided in 1965 to set up a new colony in the Indian Ocean consisting of the largely uninhabited Chagos archipelago and the islands of Aldabra, Farquhar and Desroches by agreement with Mauritius, which owns the Chagos archipelago, and with the Seychelles, which own the remaining three islands.

These developments throw very useful light on the strategic importance of the Indian Ocean and serve to give an indication of the kinds of political entanglements with the big powers with vital interests in the area which the members of the Organisation of African Unity can expect now that all the principal islands in the western part of the Indian Ocean have achieved their independence and have been admitted to membership of the Organisation.

12

The Organisation of African Unity and the future prospects of the new Africa

The outstanding fact which emerges from this study is that Africa today is a continent in the throes of rapid change and upheaval and consequently nothing that is said about the present is likely to remain valid for any great length of time. The most that one can hope to do in a brief survey such as this is therefore to focus attention on the more significant aspects of the contemporary situation and attempt to place them in the context of what has happened in the past and of what is likely to follow in the future. Our main focus of attention is the present, but we are convinced that any attempt to identify and examine the problems arising from the interactions of geography and politics in Africa today has to take account of the past, just as future developments and trends can only be discussed intelligently if they are based on the present.

There is no disputing the fact that truly profound changes have taken place in the African political scene in the course of the past quarter century, accompanied in many cases by equally striking economic developments. This vast continent which at the end of the Second World War was almost entirely under European rule has within less than a single generation succeeded in regaining its political freedom except for a few territories, mostly in Southern Africa, that are still struggling to achieve their freedom from foreign domination (Fig. 33).

Although the achievement of independence came about largely through the efforts of the individual territories themselves, these efforts were greatly assisted by the feeling of solidarity and unity in a common cause which began to emerge in the continent in the late 1950s shortly after Ghana's attainment of independence. By the early 1960s this had developed into a formidable liberation movement embracing the whole continent and dedicated to the total liberation of all African countries. The results were quite dramatic; within a few years, with the sole exception of the Portuguese and Spanish colonies, nearly all the countries of Lowland Africa had achieved their independence, while in Highland Africa the British East African and Central African territories, where the presence of large numbers of white settlers had made the prospects for independence for the African populations so uncertain and remote, were also beginning to follow suit (Fig. 32).

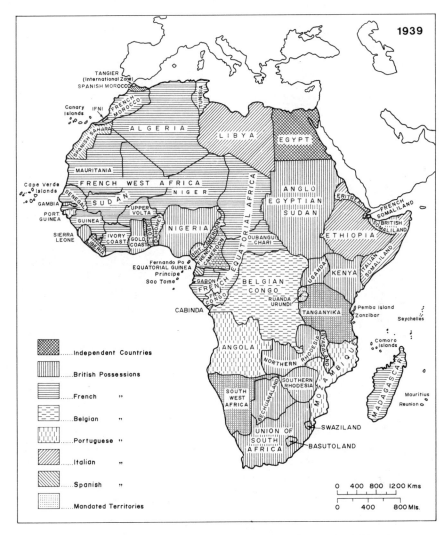

Fig. 31. Colonial Possessions and Independent States in Africa in 1939

THE ORGANISATION OF AFRICAN UNITY AND THE
AFRICAN INDEPENDENCE MOVEMENT

It was at this stage that another important development with far-reaching consequences for the future of the continent took place on the African political scene. This was the creation of the Organisation of African Unity (OAU) founded in Addis Ababa in 1963 largely at the instance of Dr Kwame Nkrumah, who had successfully led Ghana to

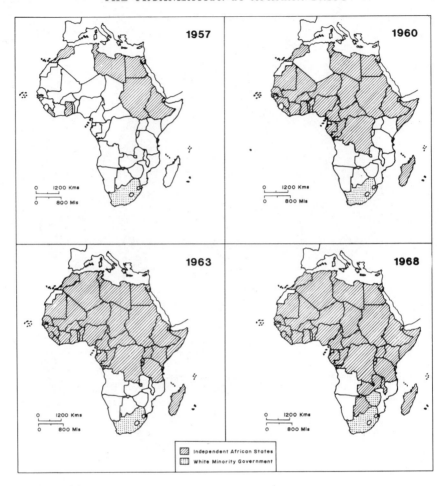

Fig. 32 Progress of Independence in Africa (1957–68)

independence in 1957, with the object of promoting unity and inter-national cooperation among African states and of eradicating all forms of colonialism from the continent. Another of Nkrumah's aims in promoting the formation of the OAU was that it should lead almost immediately to the establishment of a 'Union Government' of all African states, but this idea failed to secure the necessary support from the thirty-two founding members of the Organisation and therefore had to be suspended at least for the time being.

Many were the sceptics both within and outside Africa who doubted at the time whether such an Organisation could survive for any length of time let alone succeed in achieving any of its stated aims. Time has proved them wrong. Not only has the Organisation succeeded in speed-

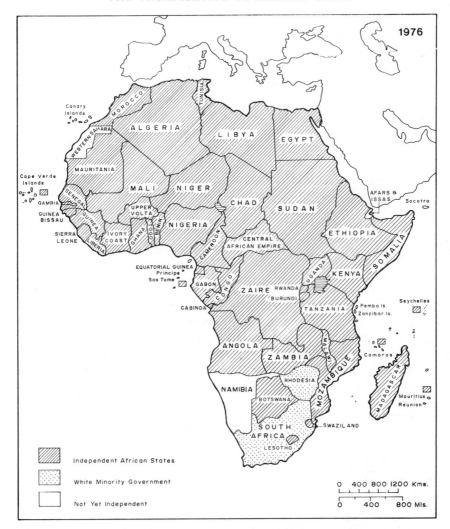

Fig. 33. Independent States of Africa in 1976

ing the political emancipation of the greater part of the continent; it has effectively presented the African viewpoint on many vital international issues. But perhaps its most significant achievement is that it has served as an important stabilising influence in the continent and assisted in the settlement or containment of many political and other disputes among its members that might otherwise have developed into serious conflicts leading possibly to war. As might be expected in view of the arbitrary manner in which most of the boundaries which now separate the new states of Africa were drawn by the colonial powers, the majority of these disputes have turned on boundary questions and the related question of

251

minorities marooned inside the territories of neighbouring states. Indeed, there are so many instances of such socio-economic anomalies due to the arbitrary imposition of boundaries in the continent that many people feared that the immediate aftermath of independence would be marked by widespread demands for boundary adjustments leading inevitably to violence and unrest. That this has not happened to any significant extent is due largely to the constructive role played by the OAU.

Another respect in which the OAU has justified its establishment is the way in which it has succeeded in maintaining a generally independent stance on important world issues and in consistently refusing to allow itself to be deflected from its main objective of protecting African interests despite all kinds of external pressures from the big powers. In this respect the Organisation has proved far more effective than any comparable organisations in other parts of the developing world.

The OAU is essentially a political organisation of equal and sovereign member states whose effectiveness in terms of actual action, like the United Nations, is limited by the important constraint that it cannot interfere in the internal affairs of any member country. This is at once a weakness and a strength – a weakness because it renders the Organisation powerless to offer assistance even where the internal actions of a member state clearly conflict with the spirit of the Organisation's Charter; a strength because by removing the Organisation from direct involvement in the internal affairs of its members it protects it from any possible charge of partisanship and therefore makes its authority more acceptable to all its members. However, in addition to its strictly political functions the Organisation also aims to coordinate and intensify the efforts of members to improve living standards in the continent as a whole and to promote international cooperation within and outside Africa, having due regard to the Charter of the United Nations and the Universal Declaration of Human Rights.

Meetings of heads of states are held once a year and are preceded by meetings of Foreign Ministers and other ministers held twice a year, usually in Addis Ababa, which is the headquarters, for the purpose of preparing agendas for the summit meetings of heads of states under a chairman elected for a one-year term of office. The day-to-day work of the Organisation is handled by a permanent Secretariat headed by a secretary-general appointed for a three-year term. There are three specialised commissions namely: (a) the Economic and Social Commission; (b) the Educational, Scientific, Cultural and Health Commission; and (c) the Defence Commission. In addition there is a Liberation Committee with its headquarters in Dar es Salaam established in 1963 for the express purpose of providing financial and military assistance to nationalist movements engaged in the struggle for independence as well as the Commission of Meditation, Conciliation and Arbitration. There is also the Scientific Council for Africa which is responsible for the coordination and direction of scientific research necessary for the promotion of economic and social development in the continent.

Since its establishment the OAU has enjoyed very close working relations with the Economic Commission for Africa, which is a United Nations Organ also based in Addis Ababa, in its non-political activities and has thus indirectly made a significant contribution to the continent's overall development. However, the principal function of the Organisation remains political, and it is chiefly in the light of its activities and achievements in this respect that its effectiveness will be judged. So far one of the Organization's main weaknesses has been the wide gap between intent and action: many lofty resolutions are frequently passed by members but few actually result in concrete action because of the lack of funds and an adequate administrative machinery for the implementation of decisions. However, it has clearly succeeded in providing the continent with an invaluable forum for the discussion of problems and issues affecting the political interests of Africa and, more especially, for the settlement of boundary and other disputes among its members. A number of serious conflicts have broken out among African countries since the establishment of the OAU, but there is no doubt that but for its mediating role the incidence and scale of such conflicts would have been very much greater. It has also served as a tremendous moral force behind the various liberation movements in those parts of the continent which are still struggling to overthrow foreign domination both by giving actual physical and military support to such movements and by actively canvassing international support for their cause through the United Nations and other world bodies. The recent withdrawal of Portugal from its African colonies, especially Angola and Mozambique, is the most notable example of the OAU's contribution to the cause of political emancipation in the continent, although the Organisation failed signally in ending the civil war which broke out among the three rival liberation movements in Angola immediately following Portugal's withdrawal from the territory and continued long after independence had been formally declared.

The OAU's conspicuous impotence in dealing with the Angolan crisis and the very dangerous threat posed by the open involvement of foreign troops and military assistance supplied by the Soviet Union, Cuba, the United States and South Africa underlines some of the deep-seated ideological differences within the Organisation. These differences have been evident since the birth of the OAU and it is important that they should not be allowed to destroy or undermine its viability as a force in African political affairs because the most serious challenges to the Organisation's strength and effectiveness are now about to begin. Now that Portuguese Africa has been liberated the independent countries of Africa will have to address themselves to the formidable task of securing the liberation of Rhodesia, Namibia and South Africa. The European minority governments which are at present in control of these territories command very considerable military and economic resources backed by a number of Western countries with important economic and strategic interests in the territories concerned, especially South Africa, and their dislodgement will call for the utmost skill in diplomacy and in the use of the relatively limited military resources available to the OAU.

The Charter of the Organisation of African Unity clearly states the commitment of member states to the Charter of the United Nations and the Universal Declaration of Human Rights. This implies the development of free, democratic institutions in all member states of the Organisation. Now that colonial rule has been overthrown in most African countries, it is clearly the responsibility of the OAU to ensure that true democratic governments are developed in its place. It is one thing for a country to free itself from foreign domination and quite another for its people to enjoy true political freedom based on the free expression of popular opinion in the management of affairs. Until this state of affairs has been established in all the independent states of Africa they cannot claim to be truly free.

AFTERMATH OF INDEPENDENCE

Since the attainment of independence the majority of African states have abandoned their initial democratic constitutions based on multiparty forms of government and replaced them with single party systems under which all opposition is banned by law. In many cases, too, civilian governments have been replaced by military governments that have come to power through military coups, and the use of the ballot box has been abandoned altogether. The reasons often given for these changes are either that the multi-party system of government is not suited to African conditions because of the internal dissensions which it invariably generates or that civilian regimes are often too corrupt to be entrusted with power; and indeed some civilian governments have so abused their power and trampled on the liberties of their subjects as to make military takeovers the only available means of salvation. But not even military regimes have proved to be stable, for in most cases they, too, have been overthrown for various reasons, including, not infrequently, charges of corruption and abuse of power, by subsequent military coups.

It cannot be denied that the practice of the Western type of democracy in many African countries has been fraught with difficulties arising from the lack of a properly informed public opinion, the high incidence of illiteracy and ignorance among the electorate and the divisive effects of tribalism, which is still a potent force in many countries. Added to these problems is the fact that most African countries still do not have sufficiently strong economic foundations to give their people enough of the basic necessities of life that are so vital for promoting that independence of outlook and sense of personal security which are essential for the true practice of democracy. But this

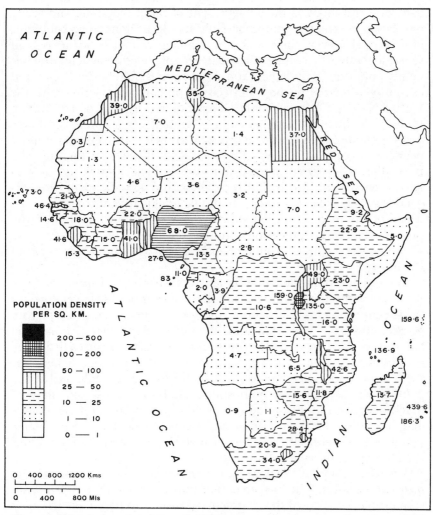

Fig. 34. Population densities of African countries in 1975 (based on UN figures)

is no reason for concluding that the concept of democracy is alien to Africa, because in fact although few of the continent's indigenous forms of government can be described as truly democratic in the Western sense many of them contain important elements and features that are democratic, however much they may differ from Western practices and forms. It should be possible, given the necessary will and imagination, for African countries to devise forms of government for their people that combine the best in their own traditional systems with the best features of Western democracy.

Without any question, however, the most urgent need in most African countries today is the achievement of rapid economic and social development; and everything possible should be done to promote this objective. However, since such development cannot be achieved and sustained unless there is a sufficient degree of popular participation it is vitally important that democratic institutions should be developed and encouraged for the purpose of decision making. In keeping with the United Nations Charter and the Charter of the Organisation of African Unity, it is the duty of all African governments to promote the active development of democratic institutions in their own countries if independence from colonial rule is to yield its full benefits for their people. Basically, as has been pointed out above, democracy is not new to Africa; what is needed is that the democratic ingredients of the indigenous systems of government should be imaginatively adapted to the needs of the modern nation states which have now supplanted the pre-colonial units of government in the continent. This is the only way whereby Africa can be assured of good as well as stable government in the future.

The big question-mark that hangs over Africa now is how soon the continent will achieve political stability and thus be able to get down to the urgent problems of economic and social development that face its people. This is not to say that no significant developments in these areas have taken place since the decolonisation process began just a little over twenty-five years ago; a great deal has in fact been accomplished, but there is no doubt that much more could have been achieved if the political climate had been more stable. Had this been so the inflow of much-needed capital from outside would have been infinitely greater and important local initiatives and efforts could have been better sustained, thus substantially increasing the pace of development and broadening its impact upon the people.

The whole approach to development has taken a new turn during the past two decades. The former notion that every country is primarily responsible for its own development according to its means has now given way to a new concept of development as an essentially international concern, and it is now much easier for the poorer developing nations to obtain development aid through such international agencies as the World Bank or through direct bilateral arrangements with developed countries and thus to strengthen their economic infrastructure.

One area where really spectacular developments have taken place is that of power and energy. Between 1937 and 1970 the production of electrical energy in Africa increased by nearly thirteen times from 6,500 million to 89,000 million kilowatt-hours per annum, with most of this growth occurring in the ten years from 1957 to 1967. The most impressive growth has been in respect of hydro-electric power. Africa's hydro-electric power potential is estimated at 145,000 megawatts or approximately 26 per cent of the world total, but so far only a little over 5 per cent of this potential has actually been harnessed. However, a growing number of countries are beginning to take advantage of their potential. Up till the 1950s very few countries produced any form of hydro-electricity, but by 1957 the number had grown to eighteen and by 1969 it was as much as thirty-one with strong indications of additional future developments. The harnessing of these resources has filled a very serious gap in the resource base of many African countries and at long last placed the possibility of large-scale industrial development within their reach. Fortunately, the possibilities for hydro-electric power generation are fairly widespread in the continent except in the very dry areas, such as the western and central Sahara, and consequently most countries are able to benefit from this source of energy. However, there are a few countries with unusually high potentials, notably Zaire (which alone is believed to possess about 56 per cent of the continent's total potential), Mozambique, Zambia and Madagascar. These countries are mostly to be found in the tropical and equatorial zones, especially where there is a combination of heavy rainfall and high relief (see p. 42).

In the area of fossil fuels, also, a number of notable or even spectacular developments have occurred since the 1950s, although these are much more localised. Until the middle 1950s Africa's known fossil fuel resources were confined to a few deposits of coal, most notably in South Africa. However, starting with the first discoveries of petroleum and gas in Algeria and Libya in the late 1950s, important reserves of these important fuels have been located in Nigeria, Gabon, Zaire and Angola, which have now joined the major oil producing countries of the world. The economic and political implications of these developments are extremely important not only for the countries concerned but for the whole of Africa, as was clearly borne out by the intense interest shown in the Angolan civil war by the great powers, which are the world's leading consumers of petroleum. If these resources are properly utilised they could bring about fundamental changes in the economies of the countries concerned and provide vital growth points for development in significant areas of the continent.

Apart from sources of energy, the post-war period has also seen very significant developments in the discovery and exploitation of valuable mineral resources, including a number of highly strategic ones, such as copper and iron. Africa has always been noted for its mineral wealth (see Fig. 2, p. 40). But the almost universal practice of exporting most of this wealth in the crude state instead of utilising it as a basis for industries located within the continent has robbed the producing countries of much of the economic returns properly due to them. Now that significant sources of power have been developed within the continent the time has come for a serious restructuring of the traditional patterns of trade so that more of the continent's natural resources, both mineral and agricultural, may be directed towards focal points within Africa and utilised internally to provide industrial support for each other within rational complexes based on African soil. The idea is not that the export of unprocessed raw materials from Africa should be excluded altogether but rather that an increasing proportion should be directed towards African markets and used for the development of local industrial centres.

An essential prerequisite of intra-African trade is the existence of an effective system of internal transport and communications and the removal of all customs and trade barriers that inhibit the free movement of people and goods across international boundaries. For a number of obvious reasons, including the long tradition of economic compartmentalisation which was fostered by the various colonial powers, these obstacles cannot be removed overnight. Even so it is clearly ironical that the inhabitants of most African countries now find it much more difficult to cross each other's boundaries than when the continent was under foreign rule. Fortunately, hopeful signs of improvement are beginning to appear, and the Trans-Africa Highway linking East, Central and West Africa, which was begun some years ago, is now nearing completion and similar projects are envisaged for the future. But the removal of customs barriers and boundary restrictions is likely to be more difficult and slow because of its economic and political implications. However, through the development of functional regional economic organisations like the East African Common Services Organisation and the more recent Economic Community of West African States (ECOWAS) inaugurated in 1975 there is hope that even these difficulties will in due course be overcome.

Closely linked to the question of internal transport and communications is the question of location in relation to external trade outlets. While intra-African trade is undoubtedly important, the economic life of practically all African countries will continue for a long time to depend very much on the extent to which they are able to trade with the rich industrialised nations outside the continent. This places a high premium on overseas trade outlets, especially railways and sea-ports, since many of the continent's exports consist of bulky raw materials which are best transported by sea rather than by air. Herein lies the importance of a coastal location and the drawbacks of location in the interior of the continent. The fourteen landlocked states of Africa are thus obliged, as has already been shown in the case of a number of these countries, to rely on those neighbouring countries with direct or easier access to the sea for their vital trade outlets to overseas markets. While such dependence necessarily compels the states concerned to maintain good relations with one another, it can cause serious hardships to the landlocked states during times of political conflict with their more fortunately placed neighbours, as has happened at various times in East and Southern Africa, though fortunately not so far in West Africa, where Mali, Upper Volta and Niger have continued since independence to enjoy reasonably good relations with the neighbouring coastal states such as Senegal, Guinea, Ivory Coast, Benin and Nigeria through which their main export outlets lie. Another serious drawback for landlocked states which has recently been underlined, particularly by the Law of the Sea Conference, is the fact that they do not have any automatic right to the valuable fisheries and other resources of the sea, especially those within the territorial waters of maritime states. The problem will become even more serious now that many coastal states are claiming 200 miles as the limit for their economic activities, unless special provision is made for them by the international community or through bilateral arrangements with adjacent or nearby maritime states.

It is nevertheless plain from the preceding chapters that the major-ity of African countries still have a long way to go before they are able to realise the full benefits of their recently acquired political inde-pendence. But, considering the continent's overall economic and human resources and the general approach of the Organisation of African Unity to the problems of development, there are reasonable grounds for optimism about the future.

One of the most serious shortcomings which the post-independence experience of many African countries has clearly underlined is the highly artificial nature of most of the continent's political divisions. As has already been shown, the pattern of states and boundaries which most African countries inherited at the time of independence was something imposed from outside and not the result of spontaneous internal evolution. Because the principal aim of colonialism was economic exploitation rather than the conscious nurturing of the col-onial territories for eventual independence as fully fledged nation states, hardly any serious attention was given to the proper integration of the diverse ethnic elements within these territories. Indeed, in a number of instances the colonial administrations deliberately encouraged internal division as a means of reducing or neutralising local opposition to their rule. The result is that many African states on the attainment of independence have found themselves faced with serious problems of internal dissension arising from tribal or ethnic differences among their people and totally lacking in that sense of unity and cohesion which is an essential prerequisite of the nation state.

It is absolutely vital that the new states of Africa should find means of strengthening their internal unity as quickly as possible so that their total human and material resources can be fully mobilised for national development and reconstruction. This is the only way in which they can effectively meet their own internal challenges and the infinitely more complex and difficult challenges of the external world, especially the problem posed by the white minority governments of Rhodesia and South Africa, which are now the only parts of the continent where the right of Africans to political self-determination has yet to be conceded.

Appendices

Basic data on African countries

Name of country	Official language	Designation of citizens	Area (sq. kilometres)	(sq. miles)	Population (1975 or latest available)	Capital	Date of independence
NORTH AFRICA							
Algeria	Arabic	Algerian	2,381,741	(919,590)	16,776,300	Algiers	3 July 1962
Egypt	Arabic	Egyptian	1,001,449	(386,666)	37,233,000	Cairo	28 February 1922
Libya	Arabic	Libyan	1,759,540	(679,359)	2,444,000	Tripoli	24 December 1951
Morocco	Arabic	Moroccan	446,550	(172,413)	17,504,000	Rabat	2 March 1956
Tunisia	Arabic	Tunisian	163,610	(63,170)	5,772,450	Tunis	20 March 1956
WEST AFRICA							
Benin	French	Beninian	112,622	(43,483)	3,112,000	Porto Novo/Cotonou	1 August 1960
Cape Verde	Portuguese	Cape Verdean	4,033	(1,557)	294,132	Mindelo/Praia	5 July 1975
The Gambia	English	Gambian	11,295	(4,361)	523,716	Banjul	18 February 1965
Ghana	English	Ghanaian	238,537	(92,100)	9,866,000	Accra	6 March 1957
Guinea	French	Guinean	245,857	(94,925)	4,416,000	Conakry	2 October 1958
Guinea Bissau	Portuguese	Guinean	36,125	(13,948)	528,000	Bissau	10 September 1974
Ivory Coast	French	Ivorian	322,463	(124,503)	4,885,000	Abidjan	7 August 1960
Liberia	English	Liberian	111,369	(43,000)	1,708,000	Monrovia	26 July 1847
Mali	French	Malian	1,240,000	(478,764)	5,697,000	Bamako	22 September 1960
Mauritania	French	Mauritanian	1,030,700	(397,953)	1,318,000	Nouakchott	28 November 1960
Niger	French	Nigeréen	1,267,000	(489,189)	4,599,785	Niamey	3 August 1960
Nigeria	English	Nigerian	923,768	(356,669)	62,925,000	Lagos	1 October 1960
Senegal	French	Senegalese	196,192	(75,750)	4,136,264	Dakar	20 August 1960
Sierra Leone	English	Sierra Leonian	71,740	(27,699)	2,983,000	Freetown	27 April 1961
Western Sahara	Arabic/Spanish	Saharaoi	265,898	(102,663)	76,425	El Aaiun	Undecided
Togo	French	Togo	56,000	(21,622)	2,222,000	Lomé	27 April 1960
Upper Volta	French	Voltaique	274,200	(105,869)	6,032,000	Ouagadougou	5 August 1960
NORTH-EAST AFRICA							
Ethiopia	Amharic/English	Ethiopian	1,221,900	(471,776)	27,946,000	Addis Ababa	Since Antiquity
Republic of Djibouti	French	Djiboutien	21,783	(8,410)	200,000	Djibouti	27 June 1977
Somalia	English/Italian	Somalian	637,657	(246,200)	3,170,000	Mogadishu	1 July 1962

263

Charter of the organisation of African unity

We, the Heads of African States and Governments assembled in the City of Addis Ababa, Ethiopia;

CONVINCED that it is the inalienable right of all people to control their own destiny;

CONSCIOUS of the fact that freedom, equality, justice and dignity are essential objectives for the achievement of the legitimate aspirations of the African peoples;

CONSCIOUS of our responsibility to harness the natural and human resources of our continent for the total advancement of our peoples in all spheres of human endeavour;

INSPIRED by a common determination to promote understanding among our peoples and cooperation among our States in response to the aspirations of our peoples for brotherhood and solidarity, in a larger unity transcending ethnic and national differences;

CONVINCED that, in order to translate this determination into a dynamic force in the cause of human progress, conditions for peace and security must be established and maintained;

DETERMINED to safeguard and consolidate the hard-won independence as well as the sovereignty and territorial integrity of our States, and to fight against neo-colonialism in all its forms;

DEDICATED to the general progress of Africa;

PERSUADED that the Charter of the United Nations and the Universal Declaration of Human Rights, to the principles of which we reaffirm our adherence, provide a solid foundation for peaceful and positive cooperation among States;

DESIROUS that all African States should henceforth unite so that the welfare and well-being of their peoples can be assured;

RESOLVED to reinforce the links between our states by establishing and strengthening common institutions;

HAVE agreed to the present Charter.

ESTABLISHMENT

Article I

1. The High Contracting Parties do by the present Charter establish an organisation to be known as the ORGANISATION OF AFRICAN UNITY.

2. The Organisation shall include the Continental African States, Madagascar and other Islands surrounding Africa.

PURPOSES

Article II

1. The Organisation shall have the following purposes:

    a. to promote the unity and solidarity of the African States;
    b. to coordinate and intensify their cooperation and efforts to achieve a better life for the peoples of Africa;
    c. to defend their sovereignty, their territorial integrity and independence;
    d. to eradicate all forms of colonialism from Africa; and
    e. to promote international cooperation, having due regard to the Charter of the United Nations and the Universal Declaration of Human Rights.

2. To these ends, the Member States shall coordinate and harmonise their general policies, especially in the following fields;

    a. political and diplomatic cooperation;
    b. economic cooperation, including transport and communications;
    c. educational and cultural cooperation;
    d. health, sanitation and nutritional cooperation;
    e. scientific and technical cooperation; and
    f. cooperation for defence and security.

PRINCIPLES

Article III

The Member States, in pursuit of the purposes stated in Article II, solemnly affirm and declare their adherence to the following principles:

1. the sovereign equality of all Member States;
2. non-interference in the internal affairs of States;
3. respect for the sovereignty and territorial integrity of each State and for its inalienable right to independent existence;
4. peaceful settlement of disputes by negotiation, mediation, conciliation or arbitration;
5. unreserved condemnation, in all its forms, of political assassination as well as of subversive activities on the part of neighbouring States or any other States;
6. absolute dedication to the total emancipation of the African territories which are still dependent;
7. affirmation of a policy of non-alignment with regard to all blocs.

MEMBERSHIP

Article IV

Each independent sovereign African State shall be entitled to become a Member of the Organisation.

RIGHTS AND DUTIES OF MEMBER STATES

Article V

All Member States shall enjoy equal rights and have equal duties.

Article VI

The Member States pledge themselves to observe scrupulously the principles enumerated in Article III of the present Charter.

INSTITUTIONS

Article VII

The Organisation shall accomplish its purposes through the following principal institutions:

1. the Assembly of Heads of State and Government;
2. the Council of Ministers;
3. the General Secretariat;
4. the Commission of Mediation, Conciliation and Arbitration.

THE ASSEMBLY OF HEADS OF STATE AND GOVERNMENT

Article VIII

The Assembly of Heads of State and Government shall be the supreme organ of the Organisation. It shall, subject to the provisions of this Charter, discuss matters of common concern to Africa with a view to coordinating and harmonising the general policy of the Organisation. It may in addition review the structure, functions and acts of all the organs and any specialised agencies which may be created in accordance with the present Charter.

Article IX

The Assembly shall be composed of the Heads of State and Government or their duly accredited representatives and it shall meet at least once a year. At the request of any Member State and on approval by a two-thirds majority of the Member States, the Assembly shall meet in extraordinary session.

Article X

1. Each Member State shall have one vote.
2. All resolutions shall be determined by a two-thirds majority of the Members of the Organisation.
3. Questions of procedure shall require a simple majority. Whether or not a question is one of procedure shall be determined by a simple majority of all Member States of the Organisation.

266

4. Two-thirds of the total membership of the Organisation shall form a quorum at any meeting of the Assembly.

Article XI

The Assembly shall have the power to determine its own rules of procedure.

THE COUNCIL OF MINISTERS

Article XII

1. The Council of Ministers shall consist of Foreign Ministers or such other Ministers as are designated by the Governments of Member States.

2. The Council of Ministers shall meet at least twice a year. When requested by any Member State and approved by two-thirds of all Member States, it shall meet in extraordinary session.

Article XIII

1. The Council of Ministers shall be responsible to the Assembly of Heads of State and Government. It shall be entrusted with the responsibility of preparing conferences of the Assembly.

2. It shall take cognisance of any matter referred to it by the Assembly. It shall be entrusted with the implementation of the decision of the Assembly of Heads of State and Government. It shall coordinate inter-African cooperation in accordance with the instructions of the Assembly and in conformity with Article II(2) of the present Charter.

Article XIV

1. Each Member State shall have one vote.

2. All resolutions shall be determined by a simple majority of the members of the Council of Ministers.

3. Two-thirds of the total membership of the Council of Ministers shall form a quorum for any meeting of the Council.

Article XV

The Council shall have the power to determine its own rules of procedure.

GENERAL SECRETARIAT

Article XVI

There shall be an Administrative Secretary-General of the Organisation, who shall be appointed by the Assembly of Heads of State and Government. The Administrative Secretary-General shall direct the affairs of the Secretariat.

Article XVII

There shall be one or more Assistant Secretaries-General of the Organisation, who shall be appointed by the Assembly of Heads of State and Government.

Article XVIII

The functions and conditions of service of the Secretary-General, of the

Assistant Secretaries-General and other employees of the Secretariat shall be governed by the provisions of this Charter and the regulations approved by the Assembly of Heads of State and Government.

1. In the performance of their duties the Administrative Secretary-General and the staff shall not seek or receive instructions from any government or from any other authority external to the Organisation. They shall refrain from any action which might reflect on their position as international officials responsible only to the Organisation.

2. Each member of the Organisation undertakes to respect the exclusive character of the responsibilities of the Administrative Secretary-General and the staff and not to seek to influence them in the discharge of their responsibilities.

COMMISSION OF MEDIATION, CONCILIATION AND ARBITRATION

Article XIX

Member States pledge to settle all disputes among themselves by peaceful means and, to this end decide to establish a Commission of Mediation, Conciliation and Arbitration, the composition of which and conditions of service shall be defined by a separate Protocol to be approved by the Assembly of Heads of State and Government. Said Protocol shall be granted as forming an integral part of the present Charter.

SPECIALISED COMMISSIONS

Article XX

The Assembly shall establish such Specialised Commissions as it may deem necessary, including the following:

1. Economics and Social Commission.
2. Educational, Scientific, Cultural and Health Commission.
3. Defence Commission.

Article XXI

Each Specialised Commission referred to in Article XX shall be composed of the Ministers concerned or other Ministers or Plenipotentiaries designated by the Governments of the Member States.

Article XXII

The functions of the Specialised Commissions shall be carried out in accordance with the provisions of the present Charter and of the regulations approved by the Council of Ministers.

THE BUDGET

Article XXIII

The budget of the Organisation prepared by the administrative Secretary-General shall be approved by the Council of Ministers. The budget shall be provided by contributions from Member States in accordance with the scale of assessment of the United Nations; provided, however, that no Member State shall be assessed an amount exceeding 20 per cent of the yearly regular budget of the Organisation. The Member States agree to pay their respective contributions regularly.

SIGNATURE AND RATIFICATION OF CHARTER

Article XXIV

1. This charter shall be open for signature to all independent sovereign African States and shall be ratified by the signatory States in accordance with their respective constitutional processes.

2. The original instrument, done, if possible in African languages, in English and French, all texts being equally authentic, shall be deposited with the Government of Ethiopia which shall transmit certified copies thereof to all independent sovereign African States.

3. Instruments of ratification shall be deposited with the Government of Ethiopia, which shall notify all signatories of each such deposit.

ENTRY INTO FORCE

Article XXV

This Charter shall enter into force immediately upon receipt by the Government of Ethiopia of the instruments of ratification from two-thirds of the signatory States.

REGISTRATION OF THE CHARTER

Article XXVI

This Charter shall, after due ratification, be registered with the Secretariat of the United Nations through the Government of Ethiopia in conformity with Article 102 of the Charter of the United Nations.

INTERPRETATION OF THE CHARTER

Article XXVII

Any question which may arise concerning the interpretation of this Charter shall be decided by a vote of two-thirds of the Assembly of Heads of State and Government of the Organisation.

ADHESION AND ACCESSION

Article XXVIII

1. Any independent sovereign African State may at any time notify the Administrative Secretary-General of its intention to adhere or accede to this Charter.

2. The Administrative Secretary-General shall, on receipt of such notification, communicate a copy of it to all the Member States. Admission shall be

decided by a simple majority of the Member States. The decision of each Member State shall be transmitted to the Administrative Secretary-General, who shall, upon receipt of the required number of votes, communicate the decision to the State concerned.

MISCELLANEOUS

Article XXIX

The working languages of the Organisation and all its institutions shall be, if possible African languages, English and French.

Article XXX

The Administrative Secretary-General may accept, on behalf of the Organisation, gifts, bequests and other donations made to the Organisation, provided that this is approved by the Council of Ministers.

Article XXXI

The Council of Ministers shall decide on the privileges and immunities to be accorded to the personnel of the Secretariat in the respective territories of the Member States.

CESSATION OF MEMBERSHIP

Article XXXII

Any State which desires to renounce its membership shall forward a written notification to the Administrative Secretary-General. At the end of one year from the date of such notification, if not withdrawn, the Charter shall cease to apply with respect to the renouncing State, which shall thereby cease to belong to the Organisation.

AMENDMENT OF THE CHARTER

Article XXXIII

This Charter may be amended or revised if any Member State makes a written request to the Administrative Secretary-General to that effect; provided, however, that the proposed amendment is not submitted to the Assembly for consideration until all the Member States have been duly notified of it and a period of one year has elapsed. Such an amendment shall not be effective unless approved by at least two-thirds of all the Member States.

IN FAITH WHEREOF, We, the Heads of African States and Governments have signed this Charter.

Done in the City of Addis Ababa, Ethiopia this 25th day of May, 1963.

ALGERIA	MALI
BURUNDI	MAURITANIA
CAMEROUN	MOROCCO
CENTRAL AFRICAN REPUBLIC	NIGER

CHAD
CONGO (Brazzaville)
CONGO (Leopoldville)
DAHOMEY
ETHIOPIA
GABON
GHANA
GUINEA
IVORY COAST
LIBERIA
LIBYA
MADAGASCAR

NIGERIA
RWANDA
SENEGAL
SIERRA LEONE
SOMALIA
SUDAN
TANGANYIKA
TOGO
TUNISIA
UGANDA
UNITED ARAB REPUBLIC
UPPER VOLTA

Bibliography

Adami, Vittorio, *National Frontiers in Relation to International Law*, trans. by Lt Col T. T. Behrens, Oxford University Press, London, 1927.

Austin, D., *West Africa and the Commonwealth*, Penguin Books, London, 1957.

Austin, Dennis, *Politics in Ghana 1946–60*, Oxford University Press, London, 1964.

Baker, S. J. K., 'Buganda, a geographical appraisal', *Transactions and Papers*, Institute of British Geographers, Vol. 22, p. 171–80.

Barbour, K. M., *The Republic of the Sudan: A Regional Geography*, University of London Press, London, 1961.

Barbour, K. M., and Prothero, R. M. (editors), *Essays on African Population*, Routledge and Kegan Paul, London, 1961.

Barbour, Nevill (editor), *A Survey of North West Africa* (The Maghrib), Oxford University Press, London, 1962.

Barros, João de, *Asia* (1552), translated and edited by Crone, G. R., Hakluyt Society, No. LXXX, 2nd series, 1937, quoted from Wolfson, Freda, *Pageant of Ghana*, Oxford University Press, 1938, p. 42.

Behr, Edward, *The Algerian Problem*, Penguin Books, Harmondsworth, 1961.

Benedict, Burton, *Mauritius*, Pall Mall Press, London, 1965.

Blake, J. W., *European Beginnings in West Africa*, Longman, 1937.

Blouet, B. W., 'Sir Halford Mackinder as British High Commissioner to South Russia, 1919–1920', *Geographical Journal*, Vol. 142, Part 2, 1976, pp. 228–36.

Boateng, E. A., *Tomorrow's Map of West Africa*, West African Affairs Pamphlet No. 13, Staples, London, 1952.

Boateng, E. A., 'Geographic Background to Current Developments in West Africa', *The Journal of Geography*, Vol. LX, No. 9, December, 1961.

Boateng, E. A., *A Geography of Ghana*, Cambridge University Press, London, 2nd edition, 1966.

Boateng, E. A., *Independence and Nation Building in Africa*, Ghana Publishing Corporation, Accra–Tema, 1973.

Boggs, S. Whittemore, *International Boundaries*, Columbia University Press, New York, 1940.

Bourret, F. M., *The Gold Coast*, Oxford University Press, London, 2nd edition, 1952.

Bovill, E. W., *The Golden Trade of the Moors*, Oxford University Press, London, 1958.

Bowman, Isaiah, *The New World*, World Book Company, New York, 4th edition, 1928.

Boyd, Andrew, and van Rensburg, Patrick, *An Atlas of African Affairs*, Methuen, London, 1962.

Brausch, Georges, *Belgian Administration in the Congo*, Oxford University Press, London, 1961.

Buchanan, Keith M. and Pugh, John C., *Land and People in Nigeria*, University of London Press, London, 1955.

Busia, K. A., *The Challenge of Africa*, Praeger, New York, 1962.

Calvocoressi, P., *South Africa and World Opinion*, Oxford University Press, London, 1961.

Carlson, Lucile, *Geography and World Politics*, Prentice-Hall, Englewood Cliffs, New Jersey, 1958.

Carr-Saunders, A. M., *World Population: Past Growth and Trends*, Oxford University Press, London, 1936.

Carter, G. M., *Independence for Africa*, Praeger, New York, 1960.

Chatterjee, S. P. (editor), *Developing Countries of the World*, National Committee for Geography, Calcutta, 1968.

Chevalier, L., *Le problème démographique nord-africaine*, Presses Universitaires de France, Paris, 1947.

Chi-Bonnardel, Regine Van, *The Atlas of Africa*, Jeune Afrique, Paris, 1973.

Church, R. J. Harrison, 'The Impact of the Outer World on Africa', in East, W. Gordon and Moodie, A. E. (editors), *The Changing World*, George Harrap, London, 1956.

Clapham, Christopher, *Haile Selassie's Government*, Longman, London, 1969.

Clarke, John I., *Population Geography*, Pergamon Press, Oxford, 1965.

Cohen, S. B., *Geography and Politics in a Divided World*, Methuen, London, 1964.

Cole, Monica, *South Africa*, Methuen, London, 1961.

Coupland, Reginald, *Indian Politics 1936–1942*, Oxford University Press, 1943.

Critchfield, Howard J., *General Climatology*, Prentice-Hall, New Jersey, 2nd edition, 1966.

Crone, G. R., *Background to Political Geography*, Museum Press, London, 1967.

Crowder, Michael, *Senegal: A Study in French Assimilation Policy*, Oxford University Press, London, 1962.

Cruickshank, B., *Eighteen Years on the Gold Coast of Africa*, Hurst and Blackett, London, 1853.

Curzon of Kedleston, Lord, *Frontiers*, The Romanes Lecture, Oxford University Press, London, 1908.

Darkwah, R. H. Kofi, *Shewa, Menelik and the Ethiopian Empire 1813–1889*, Heinemann, London, 1975.

Davidson, Basil, *Black Mother*, Victor Gollancz, London, 1961.

de Beer, Z. J., *Multi-Racial South Africa: The Reconciliation of Forces*, Oxford University Press, London, 1961.

de Kiewiet, C. W., *A History of South Africa*, Oxford University Press, London, 1941.

Demangeon, A., 'Géographie Politique', *Annales de Géographie*, Tome XLI, 1932.

Demangeon, A., *Problèmes de Géographie Humaine*, Librairie Armand Colin, 3rd edition, Paris, 1947.

Douglas Jackson, W. A., *Politics and Geographic Relationships*, Prentice-Hall, Englewood Cliffs, New Jersey, 1964.

Doveton, D. M., *The Human Geography of Swaziland*, Philip, London, 1937 (issued as a special publication of the Institute of British Geographers).

Duffy, James, *Portuguese Africa*, Harvard University Press, Cambridge, Massachusetts, 1959.

Duffy, James, *Portugal in Africa*, Penguin Books, Harmondsworth, 1962.

Duffy, James, *Portugal's African Territories: Present Realities*, Carnegie, Endowment for International Peace, Occasional Paper No. 1, New York, 1962.

East, W. G., 'The Nature of Political Geography', *Politics*, Vol. 2, 1936–37, pp. 259–86.

East, W. G., *Mediterranean Problems*, Thomas Nelson, London, 1940.

East, W. G. and Moodie, A. E. (editors), *The Changing World*, George G. Harrap, London, 1956.

Egerton, F. Clement C., *Angola in Perspective*, Routledge and Kegan Paul, London, 1957.

Emerson, Rupert, *From Empire to Nation*, Harvard University Press, Cambridge, Massachusetts, 1960.

Europa Publications, *Africa South of the Sahara*, 1971–77.

Fage, J. D., *An Atlas of African History*, Edward Arnold, 1958.

Fage, J. D., *An Introduction to the History of West Africa*, Cambridge University Press, London, 2nd edition, 1959.

Fage, J. D., *Ghana, A Historical Interpretation*, University of Wisconsin Press, Madison, 1961.

Fawcett, C. B., *Frontiers, a Study in Political Geography*, Oxford University Press, London, 1918.

Febvre, Lucien, *A Geographical Introduction to History*, Kegan Paul, London, 1932.

Fitzgerald, Walter, *The New Europe*, Methuen, London, 1945.

Fitzgerald, W., *Africa: A Social, Economic and Political Geography of its Major Regions*, 7th edition, Methuen, London, 1949.

Fordham, Paul, *The Geography of African Affairs*, Penguin Books, Harmondsworth, 1965.

Fortes, M. and Evans-Pritchard (editors), *African Political Systems*, Oxford University Press, 1940.

Frankel, S. Herbert, *Capital Investment in Africa*, Oxford University Press, London, 1938.

Furnivall, J. S., *Netherlands India*, Cambridge University Press, 1939.

Gann, Lewis H. and Duignan, Peter, *White Settlers in Tropical Africa*, Penguin Books, Harmondsworth, 1962.

Gardiner, Robert, *A World of Peoples*, BBC Reith Lectures, 1965.

Gillman, C., 'White Colonisation in East Africa', *Geographical Review*, Vol. 32, No. 4, 1942, pp. 585–97.

Gottmann, Jean A., 'Geography and International Relations', *World Politics*, Vol. 2, 1936–37, pp. 259–86.

Gottmann, Jean A., *A Geography of Europe*, George G. Harrap, London, 1951.

Gourou, Pierre (translated by Laborde, E. D.), *The Tropical World*, Longman, London, 1941.

Green, Kirk, *Crisis and Conflict in Nigeria*, Oxford University Press, 1971.

Greenfield, Richard, *Ethiopia, A New Political History*, Pall Mall Press, London, 2nd impression, revised, 1967.

Grove, A. T., *Africa South of the Sahara*, Oxford University Press, London, 1967.

Hailey, Lord, *An African Survey*, revised 1956, Oxford University Press.

Hance, William A., *The Geography of Modern Africa*, Columbia University Press, New York, 1964.

Hancock, W. K., *Survey of British Commonwealth Affairs*, Vol. 2, Part II, Oxford University Press, 1942.

Harrison Church, R. J., *Modern Colonization*, Hutchinson's University Library, London, 1951.

Harrison Church, R. J., 'African Boundaries' (Chapter XXXI) in East, W. Gordon, and Moodie, A. E. (editors), *The Changing World*, George G. Harrap, London, 1956.

Harrison Church, R. J., *West Africa*, Longman, 3rd edition, 1961.

Harrison Church, R. J., Clarke, John I., Clarke, P. J. H. and Henderson, H. J. R., *Africa and the Islands*, Longman, London, 1964.

Hartshorne, Richard, 'Recent Developments in Political Geography', *American Political Science Review*, Vol. 29, 1935, pp. 784–804; 943–66.

Hartshorne, Richard, 'The Politico-Geographic Pattern of the World', *Annals of the American Academy of Political and Social Science*, Vol. 218, 1941, pp. 45–57.

Hartshorne, Richard, 'Political Geography' in James, Preston, and Jones, Clarence, F. (editors), *American Geography: Inventory and Prospect*, Syracuse University Press, 1954.

Hartshorne, Richard, *Perspective on the Nature of Geography*, Rand McNally and Company, Chicago, 1959.

Hartshorne, Richard, 'Political Geography in the Modern World', *Journal of Conflict Resolution*, Vol. 4, 1960, pp. 55–6.

Hatch, John, *Africa Today and Tomorrow*, Praeger, New York, 1960.

HMSO, *Indians in Kenya* (Devonshire White Paper) Cmd 1922, HMSO, 1923, p. 9.

HMSO, *Report of the East Africa Commission*, Cmd 2387, HMSO, London, 1925.

HMSO, *Report of the Commission on Closer Union of the Dependencies in Eastern and Central Africa*, Cmd 3234, HMSO, London, 1929.

HMSO, *Rhodesia–Nyasaland Royal Commission Report* (Chairman: Lord Bledisloe), Cmd 5949, HMSO, London, 1939.

HMSO, *East African (High Commission) Order in Council*, HMSO, London, 1947.

HMSO, *The British Territories in East and Central Africa, 1945–50*, Cmd 7987, HMSO, London, 1950.

HMSO, *Central African Territories* (report of conference on closer association, March 1951), Cmd 8233, HMSO, London, 1951.

HMSO, *Report by the conference on federation held in London in January, 1953 (Southern Rhodesia, Northern Rhodesia, and Nyasaland)*, HMSO, London, 1953.

HMSO, *East Africa Royal Commission 1953–55 Report*, Cmd 9475, HMSO, London, 1955.

HMSO, *Report of the Advisory Commission on the Review of the Constitution of Rhodesia and Nyasaland*, Cmd 1148 HMSO, London, 1960.

Hodder, B. W. and Harris, D. R. (editors), *Africa in Transition*, Methuen, London, 1967.

Hodgkin, Thomas, *Nationalism in Colonial Africa*, Frederick Muller, London, 1956.

Hodgson, Robert D. and Stoneman, Evelyn A., *The Changing Map of Africa*, D. Van Nostrand Company, Inc., Princeton, New York, 1963.

Holdich, Sir Thomas H., *The Countries of the King's Award*, Macmillan, London, 1904.

Holdich, Sir Thomas H., *Political Frontiers and Boundary Making*, Macmillan, London 1916.

Hollingsworth, Lawrence W., *The Asians in East Africa*, Macmillan, London, 1960.

Hurst, H. E., *The Nile*, Constable, London, revised edition, 1957.

Huxley, Julian, *Race in Europe*, Oxford Pamphlets on World Affairs, No. 5, Oxford University Press, London, 1939.

International Bank for Reconstruction and Development, *The Economy of the Trust Territory of Somalia*, Washington, 1957.

Italy, Ministry of Foreign Affairs, *Report of the Italian Government to the General Assembly on the Trust Territory of Somalia*, Rome, 1960.

Jones, C. F. and Darkenwald, G. G., *Economic Geography*, Macmillan, New York, 1941, reprinted 1947.

Jones, Stephen B., 'A Unified Field Theory of Political Geography', *Annals of the Association of American Geographers*, Vol. 44, 1954, pp. 111–23.

Kay, George, *A Social Geography of Zambia*, University of London Press, London, 1967.

Kimble, David, *A Political History of Ghana 1850–1928*, Oxford University Press, London, 1963.

Kimble, George H. T., *The World's Open Spaces*, Thomas Nelson, London, 1939.

Kimble, George H. T., *Tropical Africa*, Twentieth Century Fund, New York, 1960.

Kingsley, Mary Henrietta, *Travels in West Africa; Congo Français, Corisco and Cameroons*, 3rd edition, Frank Cass, London, 1965.

Kish, G., 'Political Geography into Geopolitics', *Geographical Review*, Vol. XXXII, 1942, pp. 632–45.

Kristof, Ladis K. D., 'The Nature of Frontiers and Boundaries', *Annals of the Association of American Geographers*, Vol. 49, 1959, pp. 269–82.

Legum, Colin (editor), *Africa: A Handbook*, Anthony Blond, 2nd edition, 1965.

Legum, Colin (editor), *Africa Contemporary Record*, 1973–74, 1974–75 and subsequent editions, Rex Collings, London, 1974, 1975, etc.

Lewis, I. M., 'The Problem of the Northern Frontier Province of Kenya', *Race*, Vol. 6, pp. 48–60.

Lewis, I. M., Chapter on 'Somalia' in *Africa South of the Sahara*, Europa Publications, London, 1971.

Lewis, W. Arthur, and others, *Attitudes to Africa*, Penguin Books, Harmondsworth, 1951.

Leys, Colin and Pratt, Cranford (editors), *A New Deal in Central Africa*, Heinemann, London, 1960.

Light, R. U., *Focus on Africa*, American Geographical Society, New York, 1941.

Little, Tom, *Modern Egypt*, Ernest Benn, London, 1967.

Lucas, Sir C. P., *A Historical Geography of the British Colonies*, Vol. III, Oxford, 3rd edition, 1913.

Lugard, Frederick Dealtry, 1st Baron, *The Dual Mandate in British Tropical Africa*, 5th edition, with a new introduction by Margery Perham, Frank Cass, London, 1965.

Mackinder, Halford, J., 'The Geographical Pivot of History', *Geographical Journal*, Vol. 23, 1904, pp. 421–44.

Mackinder, Halford, J., *Democratic Ideals and Reality*, Constable, London, 1919.

Mackinder, Halford J., 'The Round World and the Winning of the Peace', *Foreign Affairs*, XXI, No. 4, July 1943, pp. 595–606.

Malthus, Thomas Robert, *First Essay on Population, 1798*, Macmillan, London, 1966.

Marsh, Zoe and Kingsnorth, G. W., *An Introduction to the History of East Africa*, Cambridge University Press, London, 1961.

Meade, James E. and others, *The Economic and Social Structure of Mauritius*, Methuen, London, 1961.

Meek, C. K., Macmillan, W. M. and Hussey, E. R. J. (editors), *Europe and West Africa*, Oxford University Press, London, 1940.

Miller, J. D. B., *The Nature of Politics*, Penguin Books, Harmondsworth, 1962.

Minter, William, *Portuguese Africa and the West*, Penguin Books, Harmondsworth, 1972.

Mitchell, Sir P. E., *Address delivered to the Central Legislative Assembly of the East Africa High Commission*, Government Printer, Nairobi, 1949.

Moodie, A. E., *Geography Behind Politics*, Hutchinson's University Library, London, 1947.

Morgan, W. T. W., 'The "White Highlands" of Kenya', *Geographical Journal*, Vol. 129, Part 2, June 1963, pp. 140–55.

Muir, Richard, *Modern Political Geography*, Macmillan, London, 1975.

Murdock, G. P., *Africa, Its Peoples and Their Culture History*, McGraw-Hill, New York, 1959.

Mutibwa, P. M., *The Malagasy and the Europeans*, Ibadan History Series, Longman, London, 1974.

Nkrumah, Kwame, *Neo-Colonialism, The Last Stage of Imperialism*, Thomas Nelson, London, 1965.

Oliver, Roland and Fage, J. D., *A Short History of Africa*, Penguin Books, Harmondsworth, 1962.

Oliver, Roland, and Mathew, Gervase, *History of East Africa*, Vol. 1, Oxford University Press, 1963.

Ominde, S. H., *Land and Population Movements in Kenya*, Heinemann, London, 1968.

Ominde, S. H. and Ejiogu, C. N. (editors), *Population Growth and Economic Development in Africa*, Heinemann, London, 1972.

Owen, Roger, *Libya, a Brief Political and Economic Survey*, Oxford University Press, London, 1961.

Perham, Margery and Curtis, L., *The Protectorates of South Africa*, Oxford University Press, 2nd edition, London, 1935.

Perham, Margery and Huxley, Elspeth, *Race and Politics in Kenya*, Oxford University Press, 2nd edition, London, 1956.

Pike, J. G. and Rimmington, G. T., *Malawi, A Geographical Study*, Oxford University Press, London, 1965.

Pollock, N. C. and Agnew Swanzie, *An Historical Geography of South Africa*, Longman, London, 1963.

Pounds, Norman J. G., *Political Geography*, McGraw-Hill, New York, 1963.

Prescott, J. R. V., 'Population Distribution in Southern Rhodesia', *Geographical Review*, October 1962, pp. 559–65.

Prescott, J. R. V., *The Geography of Frontiers and Boundaries*, Hutchinson's University Library, London, 1965.

Price, A. Grenfell, *White Settlers in the Tropics*, American Geographical Society, New York, 1939.

Ratzel, Friedrich, *Politische Geographie*, R. Oldenbourg Verlag, Munich, 1897.

Richard-Molard, Jacques, *Afrique Occidentale Française*, Editions Berger-Levrault, Paris, 1949.

Robinson, K. E., 'Political Development in French West Africa', in Stillman, Calvin (editor), *Africa in the Modern World*, University of Chicago Press, Chicago, 1955.

Sampson, Anthony, *South Africa: Two Views of Separate Development*, Oxford University Press, London, 1960.

Sauer, Carl O., *Agricultural Origins and Dispersals*, American Geographical Society, New York, 1952.

Seligman, C. G., *Races of Africa*, Thornton Butterworth, London, 1st revised edition, 1939.

Slade, Ruth, *The Belgian Congo*, Oxford University Press, London, 1961.

Stamp, L. Dudley, *Africa: A Study in Tropical Development*, Wiley, New York, 1953.

Steel, Robert, W. and Steel, Eileen M., *Africa*, Longman, London, 1974.

Stratton, Arthur, *The Great Island*, Macmillan, London, 1965.

Strong, C. F., *Modern Political Constitutions*, Sidgwick and Jackson, London, 1952.

Suliman, Ali Ahmed, Chapter on the Sudan in *Africa South of the Sahara*, Europa Publications, London, 1971.

Taylor, E. G. R., *Geography of an Air Age*, Royal Institute of International Affairs, London, 1945.

Taylor, Griffith (editor), *Geography in the Twentieth Century*, Methuen, London, 1957.

Thompson, Virginia and Adloff, Richard, *French West Africa*, Stanford University Press, Stanford, 1958.

Thompson, Virginia and Adloff, Richard, *The Emerging States of French Equatorial Africa*, Stanford University Press, Stanford, 1960.

Titmus, Richard and Abel-Smith, Brian, *Social Policies and Population Growth in Mauritius*, Methuen, London, 1960.

Udo, Reuben K., *Geographical Regions of Nigeria*, Heinemann, London, 1970.

Unesco, *A Review of the Natural Resources of the African Continent*, Paris, 1963.

Van Valkenberg, Samuel and Stotz, Carl, L., *Elements of Political Geography*, 2nd edition, Prentice-Hall, New York, 1954.

Vidal de la Blache, P., *Principles of Human Geography* (edited by Emmanuel De Martonne and translated by M. T. Bingham), Constable, London, 1926.

Walker, E. A., *Colonies*, Cambridge University Press, London, 1945.

Walker, E. A., *A History of Southern Africa*, Longman, London, 3rd edition, 1957.

Wanklyn, Harriet, *Friedrich Ratzel: A Biographical Memoir and Bibliography*, Cambridge University Press, London, 1961.

Ward, Barbara, *The International Share-Out*, Thomas Nelson, London, 1938.

Weigert, H. W., *Generals and Geographers*, Oxford University Press, New York, 1942.

Weigert, H. W., Stefanson, V., and Harrison, R. E. (editors), *New Compass of the World*, George G. Harrap, London, 1949.

Wellington, J. H., *Southern Africa*, Vols. I and II, Cambridge University Press, London, 1955.

Welsh, Anne (editor), *Africa South of the Sahara*, Oxford University Press, London, 1951.

Whittlesey, D., 'Reshaping the map of West Africa', in *Geographic Aspects of International Relations*, edited by Colby, C. C., University of Chicago Press, Chicago, 1938.

Whittlesey, Derwent, *The Earth and the State*, Henry Holt, New York, 1944.

Wieschoff, Heinrich Albert, *Colonial Policies in Africa*, University of Pennsylvania Press, 1944.

Wood, Susan, *Kenya: The Tensions of Progress*, Oxford University Press, 2nd edition, London, 1962.

Wooldridge, S. W. and East, W. G., *The Spirit and Purpose of Geography*, Hutchinson's University Library, London, 1951.

Zimmermann, Erich W., *World Resources and Industries*, Harper and Brothers, New York, revised edition, 1951.

Index